GLOBAL COMPETITION AND EU ENVIRONMENTAL POLICY

Can the EU be both green and economically competitive?

Within the EU, as in other parts of the world, concern for maintaining economic competitiveness has become a dominant political issue and, at both national and EU level, efforts are underway to streamline existing legislation. More than ever before, both national governments and the Commission are faced with considerable tension between established commitment to environmental protection and their sensitivity to promoting economic performance.

Global Competition and EU Environmental Policy is the first book to examine the relationship between economic competitiveness and environmental protection in European Union policy. A team of academics and practitioners from the fields of law, political science and economics analyse how this fear of a competitive disadvantage from high environmental regulatory standards has shaped the development of EU environmental policy. Exploring policy options which have been used in the past to reconcile competitiveness with environmental protection, the authors examine options which can offer a viable foundation for constructing the next generation of EU environmental policies. A wide range of international case studies addresses key agreements and policies, including those dealing with ozone layer protection, pesticide exports, shipping, climate change, agriculture, development assistance, and the environmental dimension of GATT/WTO.

Jonathan Golub is a Lecturer in Politics at Reading University and former Research Fellow at the European University Institute, Florence, Italy.

ROUTLEDGE/EUI ENVIRONMENTAL POLICY SERIES

DEREGULATION IN THE EU
Ute Collier

GLOBAL COMPETITION AND EU ENVIRONMENTAL POLICY
Jonathan Golub

NEW INSTRUMENTS FOR ENVIRONMENTAL POLICY IN THE EU
Jonathan Golub

GLOBAL COMPETITION AND EU ENVIRONMENTAL POLICY

Edited by Jonathan Golub

London and New York

First published 1998
by Routledge
11 New Fetter Lane, London EC4P 4EE

Simultaneously published in the USA and Canada
by Routledge
29 West 35th Street, New York, NY 10001

Typeset in Garamond by Routledge
Printed and bound in Great Britain by TJ International,
Padstow, Cornwall

British Library Cataloguing in Publication Data
A catalogue record for this book is available from the British Library

Library of Congress Cataloging in Publication Data
Golub, Jonathan.
Global competition and EU environmental policy /
Jonathan Golub.
Includes bibliographical references and index.
1. Environmental policy–Europe. 2. European University Institute.
I. Title.
GE190.E85G65   199897–38040
                                                363.7'-5'094–dc21CIP

ISBN 0–415–15698–X

CONTENTS

ILLUSTRATIONS

Figure

Tables

CONTRIBUTORS

Jonathan Golub is a Lecturer in Politics at Reading University, and former Research Fellow at the Robert Schuman Centre, European University Institute, Florence, Italy. He has published widely on topics of environmental policy, EU integration, and the interaction of national courts with the European Court of Justice. He is author of *Hard Bargains: Britain, the Environment and European Integration* (Pinter, forthcoming).

Ian Rowlands is a Lecturer in International Relations and Development Studies at the London School of Economics and Political Science. He is author of *The Politics of Global Atmospheric Change* (Manchester University Press, 1995) and co-editor of *Global Environmental Change and International Relations* (Macmillan, 1992) and has contributed articles on various aspects of environmental policy to many journals and books.

Marc Pallemaerts is a Lecturer in International Environmental Policy and Law at the Vrije Universiteit Brussel and Université Libre de Bruxelles, and legal adviser to the Belgian Secretary of State for the Environment.

André Nollkaemper is a Senior Research Fellow at the Faculty of Law of Erasmus University, Rotterdam. He has published widely in the field of international and EC environmental law.

Thomas Heller is a Professor of Law at Stanford Law School and Professor at the Stanford Institute for International Studies. He has authored numerous articles in the fields of international environmental law and policy, taxation, economic integration and labour migration.

David Vogel is a Professor of Business and Public Policy at the Haas School of Business at the University of California, Berkeley. He has written extensively on environmental policy and is the author of *Trading Up: Environmental and Consumer Regulation in a Global Economy* (Harvard University Press, 1995).

Andrea Lenschow is a Lecturer at the University of Salzburg, and former Jean Monnet Fellow at the Robert Schuman Centre, European University Institute, Florence, Italy. She has published on environmental integration in sectoral policies within the EU.

Nick Robins is European Programme Coordinator at the International Institute for Environment and Development. He is co-author of *Unlocking the Trade Opportunity* (IIED, forthcoming) which profiles positive case studies linking trade and environment.

PREFACE

This book is the first to offer a sustained analysis of the relationship between economic competitiveness and environmental protection in European Union policy. Within the EU, as in other parts of the world, concern for maintaining economic competitiveness has become a dominant political issue. At both the member state and supranational level, efforts are underway to streamline existing legislation and generally lighten what industry perceives to be excessive regulatory burdens. More than ever, national governments and the Commission are each faced with considerable tension between their established commitment to environmental protection and their sensitivity to promoting economic performance.

In light of this tension, this book brings together experts from the fields of law, political science and economics, academics as well as practitioners, to explore two central questions: first, to what extent and in what form has the fear of a competitive disadvantage due to high environmental regulatory standards in Europe shaped the development of EU environmental policy; second, which policy options have been used in the past to reconcile competitiveness with environmental protection, and do these options offer a viable foundation for constructing the next generation of EU environmental policies? The introductory chapter provides a framework for analysing the competitiveness issue, presents an overview of the individual case studies, and highlights a number of important considerations which complicate the task of accommodating both environmental and economic objectives. The seven chapters which follow address a wide range of international environmental agreements and internal EU environmental policies which have a substantial international dimension, including those dealing with ozone layer protection, pesticide exports, shipping, climate change, agriculture, development assistance, and the environmental dimension of GATT/WTO. Of central concern to each author is the compatibility of EU competitiveness and environmental protection: whether one objective undermines the other, under what conditions do environmental standards actually contribute to competitiveness (the 'win-win' hypothesis), and what are the associated implications for maintaining free trade and averting green protectionism.

PREFACE

The contributions to this volume are revised and extended versions of papers originally presented at the 1996 summer environmental workshop held by the Robert Schuman Centre of the European University Institute in Florence, Italy. The workshop and the resulting book would not have been possible without generous financial support from DGXI in the European Commission. The book is intended as a central text for advanced undergraduate and graduate courses on Environmental Politics, and is also designed for use in courses addressing Globalisation, EU Policy and Politics, and International Political Economy. Researchers and policymakers will appreciate the detail and references provided in each chapter, and will find this volume a valuable companion to the other books in the EUI Environmental Policy Series, *New Instruments for Environmental Policy in the EU* (J. Golub, ed.) and *Deregulation in the EU: Environmental Perspectives* (U. Collier, ed.).

Jonathan Golub
Florence, August 1997

1

GLOBAL COMPETITION AND EU ENVIRONMENTAL POLICY

Introduction and overview

Jonathan Golub

Born in 1973 of dual Community objectives – providing public goods while simultaneously completing the common market – there is widespread agreement that the subsequent development of EC environmental policy represents one of the most successful areas of European integration. Throughout the past twenty-five years the European Community has developed an impressive array of environmental regulations, and established itself as a main player in international environmental negotiations (Vogel 1993, Sbragia 1996a, 1996b). Community commitment to environmental policy has periodically been reaffirmed and strengthened in five Environmental Action Programmes, blueprints for the over 300 directives and regulations now in force. A similar expansion of activity has taken place in the international arena, with the Community now a party to several dozen environmental agreements.

Despite its steady expansion, however, the future of EC (now EU) environmental policy remains very much in question (Collier and Golub 1997). As in other parts of the world, an increasing sensitivity to global economic competition and budgetary constraints has made European governments wary of any form of regulation which might threaten economic growth, foreign investment, export markets, and employment creation. These pressures have been evident even in Germany, the member state most responsible since 1973 for driving up EU air, water and vehicle standards. The enormous financial costs of reunification and the emergence of the 'location Germany' debate have scuttled a number of ambitious national environmental proposals and eroded previous German environmental leadership at the EU level (Cremer and Fisahn 1998). National officials from other member states have expressed similar malaise – the much publicised 1995 Molitor report of independent experts, dominated by industrial representatives and ministers, complained of environmental over-regulation and cautioned against EU environmental policy which might jeopardise Europe's economic competitiveness (EC 1995a). Although full-scale retrenchment of environmental regulation is currently an option under discussion rather than an inevitable outcome or a *fait accompli*, the number of proposals for new EU environmental policies has slackened considerably since 1992.

Rather than a sudden obsession, however, the issue of economic competitiveness represents a perennial feature within EC environmental policymaking. On one hand, a desire to curb pollution both at home and abroad has guided Community action, catalysed in many cases by the ambitious environmental efforts of a few member states (Haigh 1984, Hildebrand 1993). On the other hand, the transmission of policy from the national to the EC level was justified in part by considerations of economic competitiveness: harmonisation of standards reflected the need to avoid green trade barriers within the Community and the perception that unregulated environmental externalities constituted an unfair source of competitive advantage for polluting states (Rehbinder and Stewart 1985). In many cases EC regulation levelled the economic playing field while simultaneously reducing pollution. But the sensitivity of regulated industries and their national governments to the high cost of pollution abatement, with its attendant effects on competitiveness, has often tempered the ability of the EC to pursue an aggressive response to environmental problems (Golub 1996b, 1996c).

A substantial literature addresses the tension between environmental improvement and economic competitiveness within the context of EC internal environmental policy, including two volumes in this series (Collier 1997, Golub 1998; see also Héritier *et al.* 1996). By contrast, relatively little sustained attention has been paid to the global competitiveness dimension of EC environmental policymaking, a topic which this volume seeks to address.[1] This chapter provides an introduction to the central issues and questions which guided this project and which are explored in the subsequent contributions. In so doing it also offers an overview of their findings and recommendations, and attempts to lay the groundwork for additional research.

The first section establishes the relevance of the global competitiveness issue within EC environmental policymaking, and suggests a framework for analysing the variety of ways in which the objectives of maintaining competitiveness and improving environmental protection conflict. It draws attention to the conflict's political salience as well as its contested empirical validity. The second section explores whether or not, and the various means by which, the global competitiveness dimension played an important role in each of the case studies. The third section then identifies a menu of possible options available for reconciling environmental and economic goals, and draws upon the case material to ascertain which approaches have predominated in EC policy. The fourth section discusses the Commission's increasing reliance upon a new approach – the pursuit of 'strategic' or 'win-win' solutions – which portrays stringent environmental policy as a means to enhance rather than undermine economic competitiveness. The serious practical and theoretical limitations of this approach are then discussed. A concluding section outlines a potential mixture of new and traditional responses to the competitiveness issue which might serve as a viable foundation for constructing and legitimating the next generation of environmental policy.

Competitiveness as an issue in EU environmental policy

Conceptualising economic competitiveness

At the most basic and commonsense level, a firm is competitive when it has the capacity to produce at costs which enable its goods to be sold at profit. But how does one conceptualise the economic performance of a single firm in relation to its competitors, or the competitiveness of entire countries or regions, or the factors which improve competitiveness or precipitate its decline? For these questions there is no single measure or definition of competitiveness. Rather, the term has come to embody a composite of a firm's production costs, productivity, export performance, capacity for technological innovation, and ability to serve and create market openings (EC 1982). Some of these factors are internal to the firm and under its control, while others, particularly the opportunity for market access, are at least partially external and affected by national as well as foreign regulations (OECD 1996). Thus it is common to think of national governments defending and promoting the competitiveness of industries within their borders, upon whose prosperity they depend for job creation, growth and revenue.[2]

The analysis of competitiveness therefore necessarily involves exploring the interdependence of government policy, individual firm performance, and national economic prosperity. In other words competitiveness depends as much on the context in which firms operate as it does on the internal design of the enterprise. The Commission has adopted an equally inclusive conception of this controversial term. According to a reflection report on the progress of the single market programme, competitiveness within the Community is generally understood to be 'the ability of firms to stand up to their competitors on domestic and world markets' (EC 1987: iii). The proper role of government and EC policy was, therefore, to help firms do this. That the famous Commission White Paper on Growth, Competitiveness and Employment failed to offer a more specific definition of individual firm, country or Community competitiveness indicates just how well established this broad and in many ways imprecise view of the concept has become (EC 1993a). In light of this, contributors to this volume were given wide latitude to explore various aspects of competitiveness, but they focus particular attention on questions of production cost, global market opportunities and international trade rules as they relate to the development of various EU environmental policies.

Competitiveness and the environment

In recent years attention has increasingly focused upon the connection between environmental policy and economic competitiveness, but the precise relationship between stringent standards and economic performance remains a matter of intense debate. Theoretically, environmental policy has the potential to function as a two-edged sword: ambitious pollution control measures may limit the ability of EU firms to compete globally, but may also be used to protect EU firms from foreign competition. Distinguishing between environmental improvements to production methods and environmental trade barriers related to product standards clarifies these opposite effects.

Conventional wisdom suggests a negative relationship between economic competitiveness and efforts to green the production process (Leonard 1988: Chapter 1, Jaffe *et al.* 1995). Regulations which require firms to reduce emissions, increase recycling, pay more for energy, or switch to more expensive fuels and ingredients all raise the final price of their products or services, with the result that green states lose markets to 'dirty' states who lack similar environmental standards. EU firms can be confronted with lower standards and thus production prices in foreign states, or can enjoy lower domestic standards than their foreign competition. Thus we would expect EU producers and service providers to oppose environmental regulation at the EU or international level which placed them at a competitive disadvantage to neighbouring producers or those based in non-EU states. The fact that national differences in production costs translate into differences in competitive position not only engenders resistance against new efforts at controlling pollution, but also creates the potential for a 'race to the bottom' or 'ecological dumping', whereby governments strategically lower their environmental standards in order to expand exports and attract foreign industrial investment.

Foreign environmental aid also has a potential competitiveness component, as it raises the question of whether EU states should shoulder the costs of environmental improvement abroad. In addition to a general reluctance on the part of states to provide substantial direct financial assistance in times of fiscal restraint, a more immediate economic concern for EU firms specialising in advanced pollution control relates to the conditions under which this technology is transferred to developing states as part of aid packages.

The link between environmental protection and economic competition is most salient when national measures threaten to disrupt international free trade by directly excluding foreign products. Market access might be denied in cases where imported goods violate domestic product standards, such as maximum emission and noise levels, prohibition of certain ingredients or requirements for a minimum amount of recycled material. Market access might also be denied to foreign goods manufactured through production processes deemed environmentally unacceptable by the importing state.

Under certain conditions, however, environmental and economic goals can be mutually reinforcing. Where penetration of cheap and 'dirty' products or services threatens to erode the market share of EU firms, whether at home or abroad, efforts by EU governments to ratchet up process standards in foreign states through international treaties and conventions can restore European competitiveness while simultaneously improving the overall commitment to global environmental protection. Negotiating international environmental agreements allows EU members to level the economic playing field and to undermine the effects of pollution havens. One would expect to find the EU advocating stringent international environmental standards and mechanisms to improve compliance in cases where it has already adopted ambitious internal policies, or where it implements current international rules more rigorously than other states. Similarly, one might expect incremental advances in EU internal environmental policy to be contingent upon reciprocity abroad.

Stringent domestic environmental regulations also reinforce the economic competitiveness of EU firms when they function as trade barriers against non-EU states, prohibiting access to EU markets by foreign products which fail to satisfy certain environmental conditions. Unilateral green trade barriers not only prevent competitive disadvantages from foreign states actively engaged in environmental dumping, but insulate EU producers against foreign goods which become comparatively less expensive as the EU raises its own standards. Because competitiveness relies so heavily on export and import penetration rates, the great risk is that nations will artificially defend their competitiveness by adopting such protectionist measures (Krugman 1994).

Herein lies the double-edged nature of the competitiveness issue: ambitious EU environmental standards might jeopardise EU firms by raising their production and compliance costs, but green EU trade barriers can insulate firms from foreign competition. In selecting a range of cases for this volume capable of illustrating both sides of the issue, attention was paid to the fact that global competition concerns enter EU environmental discourse through two routes: during direct negotiation and implementation of international agreements between EU and non-EU states, and during the formation of EU environmental legislation which significantly impacts upon economic relations between EU and non-EU states. Chapters therefore focus not only on the conclusion of international agreements but also the adoption of EU internal environmental regulations which might exceed standards set by such agreements, EU environmental regulation in sectors with substantial export markets, and provision of EU environmental aid to developing countries.

States with high environmental standards often disagree with each other when seeking to coordinate environmental laws which might raise production costs, and when determining whether national environmental policies constitute trade barriers. Within the context of strictly internal EU policy-making, disputes have erupted over the competitive effects of requiring installation of best available technology, and of pursuing environmental quality objectives rather than emission standards (Haigh 1989, Héritier *et al.* 1996, Sbragia 1996b, Golub 1997). Intense political and legal battles have also been waged over the trade-distorting effects of Danish bottle recycling programmes, stringent German car emission standards, and most recently the proliferation of national ecolabels and other new types of environmental instruments such as ecotaxes. As mentioned above, harmonisation of environmental standards at the Community level serves to avoid or eliminate the competitive effects of these and other diverging national policies.

Disagreements over what constitutes a 'dirty' product or production method, and the competitive implications of prohibiting them, also arise between the EU and other states with highly developed environmental policies, particularly in the context of global free trade. In a statement identifying the frequent point of tension, US ambassador to the EU Stuart Eizenstat noted that 'Just because the US process is different does not mean that it is environmentally unsound' (see Chapter 6). The famous US tuna embargo and many other examples in this volume illustrate the difficulty of reconciling conflicting national environmental policies when well established harmonisation or coordination procedures analogous to EU internal policymaking do not exist at the international level. Redressing the variation of environmental standards between EU and non-EU states raises even more difficult questions of economic competition because it involves potentially enormous disparities in production costs in addition to the legality of green trade barriers.

The ambiguous empirical record

A substantial literature has developed questioning the empirical validity of the presumed trade-off between competitiveness and environmental protection. If such a trade-off existed, one would expect to find signs of economic decline in countries with the most stringent environmental standards, particularly the United States, Germany, and Japan. Predictable shifts in trade, growth and investment patterns should also appear within the US, as environmental standards and their enforcement vary amongst individual states. In fact, many studies suggest that such effects are minimal or entirely absent, finding that for most industries the expenditure required to meet environmental regulations constitutes an insignificant proportion of overall production costs, often only 1 per cent; labour costs by contrast often comprise 25 per cent or more. These studies further suggest that green countries do not incur economic losses from decreased trade or industrial flight, nor do dirty countries gain competitive trade advantages and attract foreign investment by acting as pollution havens (Jaffe *et al.* 1995 provide a comprehensive literature review; see also Pearson 1987, Leonard 1988, Dean 1992). The supposedly negative relationship between environmental protection and economic competitiveness has also not been found by specific analysis of US–Mexican relations (Grossman and Kreuger 1993), US inter-state trade and investment flows (Bartik 1988, McConnell and Schwab 1990, Jaffe *et al.* 1995: 148–50), and Japanese economic performance (Nishijima 1993: 104).

Furthermore, some studies, in particular those focusing on Eastern Europe and the former Soviet Union, suggest that low environmental standards actually deter foreign direct investment rather than attract foreign industry, as environmental neglect often reflects deeper social problems and insta-bility, and is frequently associated with a lack of important factors such as infrastructure and communication which contribute much more to overall production costs (Zamparutti and Klavens 1993: 125).

However, the literature on the environment–competitiveness issue is anything but conclusive. A number of studies, even some of the ones mentioned above when read closely, note that for some industries the proportion of total production or investment costs stemming from environ-mental regulation is actually much higher than suggested by aggregates of national economic trends. In the energy, paper, oil, steel, copper and chem-ical sectors, for example, anywhere from 5–20 per cent of overall costs, and an even higher proportion of investment costs, are devoted to environmental protection. Such figures have been found for both European and American firms (Blazejczak 1993: 110, French 1993: 30, Stewart 1993: 2063–2065, Jaffe *et al.* 1995: 142). In these cases the negative effects of stringent regula-tion on product price and exports appear more significant. Studies also raise important questions about even small additions to production cost, noting the potential implications for economic competitiveness where profit margins are very tight; 3 per cent at the margin can be crucial in some industries, as even a WWF representative has conceded (Arden-Clarke 1993: 152, Jaffe *et al.* 1995: 143).

To take one example where marginal costs might have been decisive, Heerings found that EC environmental regulation had accelerated the demise of the phosphate fertiliser industry in Western Europe since 1988. As the EC adopted ever stricter discharging standards which raised domestic production costs, imports of cheap fertilisers from Northern Africa soared and EC industry relocated to developing countries (Heerings 1993: 114–118).[3]

A careful examination of the empirical record, including the material often relegated to footnotes in sanguine literature reviews (Jaffe *et al.* 1995:

152), also reveals evidence that rising environmental standards have adversely affected American and European productivity growth in the 1970s and 1980s respectively (Stewart 1993: 2073, Conrad and Wastl 1995). In the US case, the percentage of overall decline attributable to environmental regulation varied by sector but remained considerable in several important areas: 10 per cent for chemicals, 30 per cent for paper, 44 per cent for electric utilities (Barbera and McConnell 1990).

Regardless of whether the available evidence proves or disproves a conclusive relationship between economic performance and environmental regulation, the important point from the perspective of this volume is that the possible or perceived loss of competitiveness constitutes as much a political as an empirical matter, and figures prominently in industry's resistance to environmental regulation (Morris 1993: 168, Zamparutti and Klavens 1993: 120). Examples from the US and Canada attest the salience of a perceived negative relationship for policymaking. Canadian food and chemical industries have protested that domestic environmental laws disadvantage them against US farmers and firms, while the Canadian paper industry has complained that US and German environmental laws regarding newsprint paper recycling and paper production constitute disguised trade barriers which have damaged Canadian exports and forced Canadian plants to relocate to the US (St.-Pierre 1993: 98–9). Similar fears in the US over the adverse economic effects of stringent domestic environmental policies prompted the Bush administration to establish the so-called 'Competitiveness Council' under the direction of Vice-President Dan Quayle, which immediately set about scrapping existing legislation. Besides pruning its own environmental policies, US negotiators sought to level the playing field by raising standards abroad. In an effort to 'eliminate or reduce competitive disadvantages resulting from differential national environmental standards and controls', the US secured provisions against pollution havens in the NAFTA agreement and sought to introduce similar language during the GATT Uruguay Round (Barrett 1993, Charnovitz 1993: 131–132).

The link between environmental protection and competitiveness has also dominated recent discussions in Europe, with calls for deregulation, scaling back expensive international environmental commitments, and improving foreign compliance with existing rules. Even as the Maastricht treaty was proclaiming Europe's renewed commitment to environmental protection, its tight convergence criteria for countries joining the single currency were squeezing governments to reduce their deficits – spending on pollution abatement and foreign environmental development aid presented attractive targets for cuts.

Moreover, the need to reduce production costs in order to survive in increasingly global markets has played as much of a role in Europe as it has in the US. Much like their North American counterparts, German and British companies have railed against the adverse effects on international competitiveness of domestic regulation in the chemical sector (Rudnick 1992). More recently, the influential but controversial Molitor Report, produced for the Commission in 1995 by a 'group of independent experts', stressed that 'Over-regulation stifles growth, reduces competitiveness and costs Europe jobs . . . [it] hampers innovation and deters both domestic and inward investment' (EC 1995a: 1). Having singled out environmental policy as one of the areas requiring 'legislative and administrative simplifications', the Report advocated a range of deregulatory measures not unlike those of Quayle's Competitiveness Council. High on the list of proposals was the search for cheaper and more flexible regulation, and a greater reliance on new types of environmental instrument such as market based mechanisms and voluntary measures, tools which have already gained prominence at both national and EU level, and which are discussed at length in the other volumes of this series (Collier 1997, Golub 1998).

The pervasive competitiveness dimension

The perception by major actors of a significant negative relationship between economic competitiveness and stringent environmental policy constitutes a recurring theme in each of the cases analysed in this volume.

Ian Rowlands suggests in Chapter 2 that EU policy for ozone layer protection has been shaped to a large extent by a perceived negative relationship between economic competitiveness and environmental regulations which restrict the use and sale of CFCs. Because EC industries accounted for one-third of total global production of CFCs in the 1970s, and certain countries such as Britain depended upon CFCs for £70 million/year in exports and 50,000 jobs, the Community was extremely sensitive to US proposals for international restrictions on CFCs. Even after the discovery of the 'ozone crater' over Antarctica in 1985 sparked renewed efforts to control CFCs and methyl bromide, both the extent and manner of their reduction were guided by concern for economic competitiveness of EC firms vis-à-vis the US, as well as by intra-EC economic battles where substantial local adjustment costs (14,000 jobs in the case of Spain), product development costs, competing CFC substitutes and export markets were at stake.

Much like the case of CFCs, the development of EU restrictions on the export of hazardous chemicals, explored in Chapter 3 by Marc Pallemaerts, posed enormous consequences for the competitiveness of EU firms. In the late 1970s, EC-based companies accounted for over one-third of world pesticide sales, and had a particular stake in export markets outside Europe where two-thirds of all their sales were concentrated. As a result, for many years EC internal regulations governing pesticides were not extended to cover exports, a position supported until 1983 by the Commission. In the mid-1980s, intra-EC economic competitiveness became a central issue in policy development, as a number of member states adopted their own restrictions on chemical exports and sought similar action at the EC and international level. Amidst intense debate within the Council, a fully-blown prior informed consent procedure which would limit EC export of chemicals was proposed in 1990, but only took meaningful effect in 1994. During this time attention was constantly focused on negotiations within the context of GATT, UNEP and the FAO, in order for the EC to avoid or compensate for expensive unilateral restrictions by imposing a legally binding PIC regime on Japan, the US, Canada, Australia and other states competing for the lucrative chemical export market.

Economic competitiveness has emerged as an increasingly important issue in the field of marine environmental protection, as André Nollkaemper demonstrates in Chapter 4. Whether transporting waste or oil, the operation of sub-standard vessels with poorly trained crews and operators increases the risk of environmental disasters. While the EC and some individual member states have steadily tightened up their own shipping laws and domestic implementation of international rules, discrepancies remain throughout the community and equivalent measures have not been applied with similar force in many non-EC states, resulting in a substantial compliance gap which damages the environment and poses considerable economic problems for the greener member states. Up to 13 per cent of a vessel's total annual running costs derive from meeting international standards, including environmental standards, with more ambitious national environmental laws imposing an additional 1–2 per cent. Flagging out, port shopping and avoidance of expensive environmental technology results, as ship owners, port operators and flag states compete for business and competitive advantages in a sector renowned for tight profit margins. The substantial decline in the number of ships flying EC flags, particularly Dutch flags, has been blamed in part on competitive disadvantages incurred by ports and flag states who enforce EU and international environmental laws. Within the EU, although no studies of the matter have been conducted, national officials have expressed concern that business might have shifted towards those member states where compliance with standards is relatively lax, most notably Greece.

Issues of economic competition have also dominated the development of an EU carbon–energy tax. As Thomas Heller discusses in Chapter 5, the fate of the Commission's 1992 proposal to reduce carbon dioxide emissions through an incremental levy on fuel hinged primarily on whether or not member states could agree on how to minimise or distribute the predicted impacts on economic growth, employment and inflation levels throughout the Community (see also Verbruggen 1993: 57). Negotiations over a possible burden-sharing solution were deadlocked, as no state proved willing to shoulder a disproportionate responsibility for achieving the Community's overall emission stabilisation target originally agreed at Rio, whether through fuel switching, investment in more efficient generation of energy, dramatic reductions in energy consumption, or substantial expenditure on environmentally friendly modes of transportation. Economic competition between the EU and its major trading partners also played a considerable role in the negotiations, as member states conditioned adoption of an EU tax on the US, Canada and Japan undertaking similar expensive measures.

The range of cases assessed in Chapter 6 by David Vogel illustrates the direct and substantial effect on economic competitiveness of green trade restrictions in the US and EC. Penalties imposed for failing to meet US car emissions standards cost European car exporters $500 million in 1991, while the 1990 and 1992 US bans on tuna caught with nets which endangered dolphins cut off exports from a variety of trading partners, including France, Italy, and Spain. Similarly, the proposed EU ban on furs caught with 'inhumane' traps would foreclose a $30 million export market, and threaten 80,000 Canadian trappers with potential economic ruin. Ecolabelling schemes do not directly exclude products, and thus are not usually classified as traditional trade barriers, but raise equally important issues of economic competition. The EU ecolabel has come under attack from the US and developing states as a disguised barrier to trade which establishes a variety of capricious criteria designed specifically to reward EU products which are not necessarily 'greener' than foreign goods.

In contrast to the other cases in this volume, Andrea Lenschow's analysis of the Common Agricultural Policy in Chapter 7 demonstrates an area where, until recently, the perceived negative effects of environmental regulation upon economic competitiveness have played only a marginal role. By providing EU farmers with guaranteed prices – import levies, export subsidies and direct intervention purchasing – the CAP sheltered its domestic producers from the pressure of international competition and alleviated some of the potential economic consequences of imposing stringent environmental controls. The lack of a green political agenda and the extreme insulation of powerful agricultural interests within the EU decision-making process – domestic rather than international factors – prevented the introduction of anything beyond modest environmental regulations. However, the relationship between environmental standards and the economic competitiveness of EU farmers has become an increasingly relevant issue in the debate over how to structure the CAP in such a way as to accommodate international trade liberalisation, EU enlargement, and the subsidiarity principle.

Economic competitiveness has also emerged as an important consideration in the area of EC aid to developing countries. As Nick Robins shows in the final chapter, throughout the history of EC aid efforts the Community has periodically reviewed its own aid programmes in light of their environmental effectiveness, incorporating sustainable development goals, environmental impact assessments and project cycle management provisions into the Lomé Convention and other aid programmes. The financial value of EC aid devoted to environmental projects has also been substantial, reaching 1 billion ECU in 1994, and constituting approximately 10 per cent of all EC development aid in the first half of the 1990s. But government efforts to foster sustainable development abroad through green financial aid and technology transfer face mounting pressure from competing economic priorities – general fiscal constraints and competition for market share amongst producers in the rapidly growing multibillion dollar market for high technology pollution abatement equipment. To take the most striking example, four years after the UN Earth Summit in Rio de Janeiro the EU has failed to deliver on its $4 billion aid commitment towards implementing global sustainable development.

Reconciling environmental protection and economic competitiveness

When faced with the threat of economic disadvantages produced by stringent environmental policies, the EU and its member states have traditionally relied upon four policy options which might be analogised to 'carrots' and 'sticks' (Charnovitz 1993: 143–144). When choosing to use 'carrots', governments alleviate perceived competitive disadvantages either by simply relaxing their domestic environmental standards, or by compensating for the effects of high standards through the use of domestic subsidies, exemptions for affected industries and other offsetting measures. In contrast, when opting for 'sticks', the EU and individual member state governments attempt to overcome competitive disadvantages by forcing other countries (including reluctant EU states) to raise their environmental standards. This can be done by pursuing more stringent international agreements and harmonised EU standards, as well as by seeking to improve compliance with existing international rules and EU legislation. When placed at a competitive disadvantage against non-EU states, the Community can also consider the possibility of offsetting the costs of its internal environmental policies with countervailing duties (CVDs), a particular type of 'stick' which places green taxes on imports (Charnovitz 1993: 142–143).

Relaxing environmental standards

Contrary to expectations that competitiveness might ignite a 'race to the bottom' and provoke widespread use of the 'carrot' approach, evidence from the case studies suggests that EU governments do not strategically repeal existing environmental legislation in order to attract investment or maintain market share for their domestic industries. Nor has the EU systematically sought to withdraw from international environmental agreements to which it is a party. In only two cases (see Chapters 4 and 8) did EU governments actually lower their environmental standards and commitments. To bolster their declining shipping industries, the governments of Germany, Denmark and Finland created 'second registers' with lower standards similar to those found in 'open registers', including less demanding requirements for crew training and equipment which helps prevent marine pollution. In this same sector, the Dutch government is considering whether to relax some of its own environmental standards in order to cut costs. The second instance involved concern for the general economic competitiveness of EU states and the perceived opportunity costs of providing large sums of foreign environmental aid. In recent years, pressures to reduce public spending have hit aid programmes hard, and in 1994 the EU reneged on its $4 billion aid pledge made at Rio. Most EU states also reduced their total levels of environmental development assistance. Nick Robins has suggested that Rio was the 'death knell of international welfarism', including green aid (Robins 1996: 10). In addition, conscious of the need to gain economic advantage from the spending that remains, foreign aid is increasingly aimed at improving investment and trade links for donor states, rather than fostering sustainable development in the recipient countries.

Domestic offsets

While an outright race to the bottom has not occurred, governments have consistently deployed a range of domestic 'carrots' to offset the perceived effects of environmental policies on economic competitiveness. Measures which alleviate the costs of current and proposed environmental regulation include: government subsidies for pollution control, financial aid for research and development, lax enforcement, and postponement or avoidance of costly new legislation which would disadvantage domestic industry. In some cases, which are discussed later in this chapter, reliance upon offsets has led to an extreme version of the 'carrot' approach, sometimes referred to as a 'no regrets' policy. This involves adopting regulation only when environmental improvements can be achieved without incurring any additional economic costs – in short, the possibility of getting a free lunch. While not entirely free, the solution advocated by the Molitor report, and one already apparent in both national and EU level legislation, involves searching for a 'cheap lunch' – minimising regulatory costs by shifting from the traditional command and control approach towards an arsenal of new policy tools such as ecotaxes, ecolabels and voluntary agreements (Collier 1997, Golub 1998).

In the case of marine pollution (Chapter 4), government offsets in both flag states and port states occur through lax implementation of EU and international rules, producing a situation much like a race to the bottom. While states in the Middle East and North Africa top the list of environmental laggards, ships flying the flags of Portugal, Greece and Denmark have the highest rates of failure within the EU, a dubious honour which

Nollkaemper attributes in part to the pressures of economic competitive-ness. The 'carrot' approach has also been tried, for example in Rotterdam, where the government subsidised the reduction of port costs for tankers using segregated ballast tanks (SBTs), an environmentally beneficial feature which owners are disinclined to invest in because it adds to a ship's size and increases its tonnage dues in port. The IMO has encouraged all states to follow similar offset policies aimed at reducing marine pollution. At a more general level, the shipping sector has received widespread state aid throughout the EU, assistance which can be used to cover the costs of meeting international and EU environmental regulations.

Pallemaerts (Chapter 3) identifies several cases in the field of pesticide exports where EC laws included attempts to offset competitive disadvan-tages. The Commission's 1986 proposal for an 'informed choice' system, which would have restricted EC exports more tightly than existing interna-tional guidelines suggested, was rejected by the Council, and the 1988 regulation eventually adopted, according to Pallemaerts, had only a minimal effect on EU trade. The prior informed consent procedure adopted in 1992 also contained provisions designed to limit its economic impact on EU producers, as enforcement was left to the discretion of the member states and chemicals actually subject to restriction were only identified in 1994.

Perhaps more than any other area covered in this volume, EU policy on climate change (Chapter 5) exemplifies governmental recourse to offsets and no regrets policies as a means of ameliorating competitive disadvantages from environmental regulation. The four member states which have enacted carbon taxes have all exempted energy intensive industry and set tax rates at extremely low levels. Similar exemptions, as well as general tax refunds, were made an integral part of the ill fated EU carbon–energy tax, negotia-tions over which nevertheless remain deadlocked mostly because of sensitivity to issues of competitiveness. It was hoped that a burden-sharing approach coupled with an unspecified form of intra-EU transfer payments would facilitate political agreement by offsetting the costs in the hardest hit member states. The EU strategy also contained research subsidies under the ALTENER programme. Finally, as was the case with individual national policies, EU proposals also took a carrot approach through 'fiscal neutrality' – offsetting energy taxes with lower employer contributions for social secu-rity and other expenses, and by earmarking the collected tax revenue and returning it to industry in the form of subsidies.

Heller suggests that joint implementation, an innovative form of offset which minimises EU expenditure by targeting emissions abatement in developing states where the marginal costs of CO_2 reduction are much lower, provides the only realistic means of preventing global warming which reconciles the EU's dual concern for competitiveness and environmental protection. The impending exponential increase in emissions from Asian states over the next fifty years could well trigger global warming regardless of whether Europe were to reduce dramatically its own use of fossil fuels (the so-called 'China Trap'). And compared to what additional sharp cuts in CO_2 would cost the already highly regulated member states, the potential emissions reductions made by altering the trajectory of Asian transportation and energy developments constitute relatively inexpensive 'low hanging fruit', offering cost savings of 70–85 per cent. But to avoid the 'China Trap', the EU and other developed countries must act quickly to help reduce Asian emissions, lest these fruits rot on the vine as Asian states commit themselves to inefficient and polluting growth paths.[4]

Domestic offsets have also been used to prevent farmers from facing the full cost of remedying environmental damage attributable to the CAP (Chapter 7). Although a longstanding goal of EC environmental policy, and an explicit provision of the 1987 Single European Act, the Polluter Pays Principle has never been applied to agricultural production, and the farm lobby has successfully resisted the few proposals which would have introduced ambitious measures to combat diffuse non-point source pollution from fertiliser. In addition to the lucrative price supports mentioned above, the handful of environmental measures adopted by 1992 provided subsidies for land set-aside and direct payments to farmers who undertook environmentally friendly projects.

Under pressure during the Uruguay GATT round to liberalise international trade and reduce traditional forms of agricultural subsidies, EU officials actively embraced offsets which effectively greened the CAP because such measures were some of the few remaining GATT-legal means of supporting EU farmers. Deprived of the ability to maintain price supports at the same level, the EU sought to compensate farmers for their losses through two measures acceptable under GATT rules – direct income support decoupled from production, and environmental payments. Capitalising on the 'selling points' of environmental measures, the EU supported creation of the 'green box' which exempted these two sources of support from GATT scrutiny, thereby offsetting a mandatory 20 per cent drop in domestic subsidies.

Evidence suggests that the partial decoupling of financial assistance from production has produced a notable drop in the use of chemical fertilisers and pesticides. However, liberalisation pressures will never adequately green the CAP, as a range of traditional as well as new considerations exert powerful constraints. Moreover, the widespread reliance on environmental offsets from Brussels as a source of aid, income and guaranteed market share risks the possibility of an intra-EU regional subsidy war if the CAP is renationalised in accordance with the subsidiarity principle, as southern farmers struggle to secure local subsidies and maintain competitiveness against better financed northern producers.

One of the most important lessons which emerges from the cases is that, while it certainly plays an important role in reconciling competitiveness and environmental protection, the absence of a 'carrot' approach has not precluded the EU or some of its respective member states from pursuing ambitious environmental objectives. One obvious reason for this is that scientific and technological advances, as well as highly publicised environmental disasters, have generated a rising baseline of environmental policy within European countries. Major political parties have also incorporated environmental themes in order to capture voters and neutralise the attractions of Green parties (Richardson and Rootes 1995, Collier and Golub 1997). However, while most states have embraced environmental rhetoric, this does not suggest a convergence in national environmental perspectives – important national differences have remained and in some cases intensified.

The primary source of green leadership has come from states with strong traditions of environmental protection, most frequently Germany and the Netherlands, often working together with the Commission. Pallemaerts draws attention to the fact that Germany, the Netherlands and the UK each pursued unilateral restrictions on chemical exports in 1985–1986. Rowlands notes that during the same period, contrary to what one might have expected based on industrial interests, Germany pushed for aggressive restrictions on CFCs. Heller describes how individual member states moved to reduce CO_2 emissions in 1990, prior to the Rio conference, and how in the same year the Council adopted an emissions stabilisation target. And Vogel discusses EU environmental trade initiatives designed to promote international protection of animals and forests. In each of these cases economic competitiveness did not prevent policy development, nor were domestic firms fully compensated through government offsets.

This certainly does not imply that competitiveness issues placed no constraints on environmental upgrading in these leader states. On the contrary, an equally important lesson from the cases is that when greening does occur in a progressive state, this is often foreshadowed by, and certainly followed by, efforts to compensate for its costs through EU-wide and then international regulation. The remainder of this section concentrates on this phenomenon.

Forcing standards up

By far the most common response within the EU when facing the environmental–competitiveness dilemma, including in the cases of environmental leadership mentioned above, has been recourse to 'sticks' rather than 'carrots'. States have advocated EU harmonisation and supported international conventions which coincide with their existing or planned domestic measures.

EU policy to combat climate change (Chapter 5) exemplified this approach when application of the carbon–energy tax was made contingent on other OECD states adopting similar measures in accordance with a 'conditionality clause'. In the case of shipping (Chapter 4), a few EU efforts to reduce discharges at sea and to improve domestic compliance with environmental rules were made after the famous *Amoco Cadiz*, *Aragon* and *Khark-V* accidents in 1978–1979, but the 1992 disaster off the Shetland Islands was the real watershed which prompted the aggressive new plans contained in the 'Common Policy on Safe Seas'. The Common Policy will greatly improve compliance within the EU, but threatens to exacerbate the competitiveness gap with non-member states. The Commission contends that greater reliance on its external competence and a united EU negotiating position in the IMO would allow it to project similar improvements abroad, thus closing the gap.

Since 1983, the Commission has tried to achieve environmental protection in the field of pesticide exports (Chapter 3) while minimising competitive losses by pegging its innovations as much as possible to international developments, for example by developing internal PIC procedures, then pushing for similar legally binding international rules. Here again we see competitiveness issues arising between the EU and its trading partners as well as amongst EU member states. The 1986 Commission PIC proposal was shaped by discussions within international arenas such as the FAO, OECD and UNEP, but it also reflected the interests of the three member states who had already adopted stringent unilateral domestic standards. Although originally blocked by the majority in the Council, when the EC eventually adopted a legally binding PIC procedure in 1990 the Commission immediately abandoned its previous reservations and sought to impose similar measures abroad through GATT, UNEP and FAO rules. In stark contrast to its 1988 position on the tuna–dolphin case, where the EC accused the US of violating GATT rules by pursuing environmental protection beyond its own borders, when facing competitive disadvantages from its own unilateral PIC regulations for exportation of pesticides, the EU explicitly endorsed the need for governments to incorporate extraterritorial environmental considerations into their trade policies.[5]

As Rowlands shows in Chapter 2, the 'stick' approach has been used consistently by governments when devising policy to protect the ozone layer, with the US seeking to raise EC standards and green member states seeking to raise and harmonise EC standards. In 1977, the US banned CFCs in aerosol cans, and then pushed for similar international action. Sensitive to the interests of its CFC producers, the EC limited its own legislation to a production cap that safeguarded domestic CFC usage as well as lucrative export markets. In the mid-1980s, several member states adopted unilateral CFC restrictions and then attempted to share the economic burden by raising EC standards.

A similar pattern emerged in the late 1980s and 1990s after scientific consensus had generated widespread pressure for an aggressive CFC phase-out: the race for CFC substitutes and export markets conditioned national negotiating positions. The UK and Germany, along with the US, supported

17

sharp HCFC reductions because they were already exploring the next generation of substitutes, whereas France resisted such reductions because it had invested heavily in HCFCs as the preferred substitute. For methyl bromide, the US again was the first to restrict its use and sought similar international measures in part because US farmers perceived a threat from Mexico and Southern Europe where economic competitiveness depended on using this crop-enhancing chemical. Of course southern EU states resisted out of fear for their competitive position, while many northern states who do not produce any methyl bromide, and had already taken or planned unilateral reductions in its use for environmental reasons, supported the US.

As noted at the start of this chapter, resolving the tension between economic competitiveness and environmental protection by forcing other states to adjust their standards upwards can take the form of green protectionism. As Vogel shows in Chapter 6, each of the EU positions taken in relation to environmental trade has served to defend EU firms against foreign environmental measures, or to shelter EU producers from foreign products. Thus the EC opposed US car emissions (CAFE) standards which disproportionately harmed European exporters. In decrying the extraterritorial effect of the US tuna bans, the EC remained sensitive to how depressed prices from a glutted market would harm its own tuna industry.

But, as mentioned above, in several areas the EU has supported extraterritorial environmental concern when this has coincided with the economic interests of its producers. The US, Canada and several developing states have attacked the EU ecolabel and various national labels for their improper use of life-cycle analysis, their unjustified emphasis on using recycled materials and their lack of transparency, features which place the EU's trading partners at a competitive disadvantage for product markets (a point also raised in Chapter 8). The Commission has retained the stick approach, attempting to ensure the legitimacy of the EU ecolabel as part of ongoing WTO reform (EC 1996a) while avoiding equivalence with less demanding ISO standards (Taschner 1998). Two more examples are the 1983 EC ban on imports of seal skins, and the 1991 ban on importation of furs caught using leg-hold traps, each of which supported the EC pelt industry. Similarly, the EU is considering restrictions on exported Malaysian wood, which would protect its own timber industry, after having strongly criticised an Indonesian timber export restriction which would have protected the Indonesian wood processing industry. In none of these cases has the EC supported or taken action which improves the environment at the expense of its own economic competitiveness.

Countervailing duties

The cases in this volume suggest that, amongst the many 'carrots' and 'sticks' available to neutralise the perceived competitive disadvantages stemming from stringent domestic environmental policies, governments only occasionally seek to impose countervailing duties. Two instances (in Chapters 2 and 5) stand out as exceptions to this general finding. Under Spanish leadership, the 1995 Vienna Conference agreed to phase out methyl bromide by 2010 but included the possibility of trade measures against non-parties (Spain was particularly concerned about Morocco) to level the playing field. The possibility of incorporating CVDs as a means of safeguarding economic competitiveness was also considered during the protracted negotiations over the EU carbon–energy tax .

While CVDs have traditionally played a minor role in EU environmental policy, they have recently emerged as an important agenda item in the ongoing debates over 'greening' the GATT/WTO (Esty 1994). The EU supports greening the GATT for a variety of factors, one of which is undoubtedly a sincere commitment to improving international environmental protection. But economic considerations, including the desire to preserve competitiveness in light of high EC standards and the fear of being swamped by cheap imports from relatively unregulated former Eastern bloc countries, have also played a major role. The debate centres more on how, not whether, to green international trade rules, and the EU position on unilateral efforts to force up foreign environmental standards remains finely balanced, if not openly hypocritical. In fact, no country is immune to charges of hypocrisy – the EC, Canada, the US and other OECD states all struggle to shape international trade rules which legitimate their own policies.

As Vogel discusses in Chapter 6, officially the EU opposes levies and other forms of environmental trade barriers (EC 1992, 1996a), although the European Parliament has advocated reforms which allow 'non-tariff trade barriers to protect the environment' (EP 1993b). The Commission has moved between these two positions depending on the issue at hand. For example, originally EC opposition to the US unilateral tuna embargo was 'a convenient stick with which to beat the Americans in the margins of the Uruguay Round' (EP 1993a: 4) but the threat of one million tonnes of tuna being dumped on EC markets generated sudden interest in saving dolphins and support for unilateral measures having extraterritorial effect. The ecolabel controversy between the EU and its trading partners exemplifies the potential for green protectionism, and the pressure for WTO reform created by a new generation of environmental policy tools which blur the product/process dichotomy underpinning traditional environmental trade law (EC 1996a).

Signs of green protectionism have also appeared in the EU's handling of CAP reform. Lenschow notes that EU ministers have expressed alarm over additional moves towards agricultural trade liberalisation which would expose them to competition from non-EU states with lower environmental standards. Unless offset by domestic subsidies, CVDs, or direct trade barriers, the EU's relatively stringent environmental policies may eventually disadvantage its farm sector when entering bilateral agricultural trade agreements (for example with South Africa). And if EU environmental standards are not maintained uniformly across the Union after its enlargement, similar competitive divisions may emerge between the present agricultural community and farmers in candidate countries.

Strategic EU environmental policy

In contrast to the traditional sticks and carrots approach which acknowledged and sought to neutralise the negative economic effects of environmental regulation, recent debates have centred around the new paradigm of 'strategic environmental policy' which questions the inherent tension between economic and environmental goals. Rather than envisaging a race to the bottom, this theory suggests that unilateral action actually improves economic competitiveness, thereby igniting a race to the top (Barrett 1993). Part of this new approach is to recast environmental costs as investments in long-term competitiveness and self-interest. The argument, increasingly heard in Europe, that environmental protection provides long-term economic advantages, echoes similar claims made by American scholars struggling to reconcile US environmental regulation with competitiveness (Porter 1990, Porter and van der Linde 1995).

The Commission has espoused this new perspective in a series of recent publications, and has grounded the legitimacy of ambitious EU supranational regulations on their ability to promote 'win-win' solutions – instances where rising environmental standards not only diminish pollution, but reduce production costs and improve competitiveness (EC 1992, 1993b, 1994, 1996a). The strategic environmental policy view argues that firms do not automatically operate at maximum efficiency and profitability because market failures, including operational and transaction costs, prevent them from automatically pursuing innovation and savings. Thus the environmental regulation so adamantly resisted by industry actually generates a range of net financial gains: by guaranteeing clean inputs for manufacturing, spurring more efficient modes of production, preventing disruptions to production caused by environmental accidents, improving workers' health and productivity, guaranteeing EU products access to foreign markets, meeting green demands from foreign consumers, and positioning EU firms with first mover advantage to capture the lucrative $200 billion global market for green technology and services.[6] Not only do regulated industries actually stand to gain financially, so the argument goes, but by reducing pollution governments can improve the business environment, thereby encouraging foreign investment and preventing capital flight.

Studies of how environmental regulation has affected the performance of firms in the US, Germany, Japan and Australia provide numerous examples of missed investment opportunities and after-the-fact savings which support Commission claims that win-win solutions abound (Blazejczak 1993: 110,

Porter and van der Linde 1995). Rather than arising spontaneously from market forces, these savings have occurred through regulation, as inertia, risk and uncertainty are overcome by levelling the playing field with tough standards (van der Linde 1993: 74–76).

The chapters in this volume suggest that the strategic approach has played a minor role compared with reliance upon traditional sticks and carrots, but that in a few instances the Commission and individual member states have built win-win provisions into their environmental policies.

In Chapter 5 on EU climate change policy, Heller discusses how the Commission has based its strategy for selling a carbon–energy tax to reluctant member states on the promise that environmental protection yields a 'double dividend' of increased growth and employment, the clearest example to date of a win-win philosophy. The Commission contends that restructuring national tax bases, deriving revenue not primarily from the usage of labour by firms as is currently done, but by taxing their consumption of environmental resources, will stimulate technological development and enhance EU competitiveness while also averting the consequences of global warming (see DRI 1994).

Robins explores in Chapter 8 how the 'second generation' of green development assistance has involved a search for win-win solutions through the use of tied aid – improving environmental conditions abroad while simultaneously increasing exports of EU services and high technology pollution abatement equipment. Officially, EU aid policy is designed in part to promote use of environmentally friendly technology by developing states (Robins 1996). However, the Community recovers 48 per cent of this money through purchases from European companies by recipient states. And 80 per cent of Lomé contracts go to EU firms. A similar win-win strategy has been adopted at the national level: Denmark and France each offer substantial amounts of aid, but in both cases 35–50 per cent is tied to purchases in the donor state and national officials extol the domestic economic objectives of green development assistance. Unfortunately, the intra-EU competition issue re-emerges here, as states are reluctant to screen or in any way hinder export promotion activities. Although eventually halted, British funding for the Pergau hydro-electric scheme in Malaysia highlights the potential environmental consequences of tied green aid.[7]

Besides reliance on tied aid, the emergence of an EU 'environmental security' dimension may offer a novel form of legitimating international environmental assistance as part of a win-win strategy. Instead of focusing on the cost of providing aid, this new approach would emphasise how sustainable development promotes the EU's economic and political security by guaranteeing a steady long-term supply of materials, improving stability in developing states, and preventing flows of environmental refugees (see Chapter 8).

In contrast to the climate change and green aid examples, Nollkaemper finds in Chapter 4 that neither the Commission nor individual EU member states have followed a win-win approach towards controlling marine pollution from the shipping sector. The Commission's reluctance to adopt a strategic perspective appears somewhat surprising in light of a 1993 analysis which predicted that stringent environmental rules and improved enforcement would undoubtedly damage competitiveness in the short run but would yield win-win results over the long run by inducing innovation. On the other hand, the steady decline of EU shipping and continued flagging out make aggressive unilateral measures appear less attractive.

Nor did a win-win philosophy underpin national or EU policies to protect the ozone layer, although some authors, including Rowlands (Chapter 2), suggest that *ex post* analysis has revealed that international restrictions might have actually yielded net savings where substitutes cost less than the CFCs they replaced. According to these accounts, while for many years producers had knowledge of these potential substitutes and in some cases were investing in their development, government regulation was a necessary prerequisite for their actual introduction and the attendant cost savings. Nevertheless, Rowlands notes that once international regulation became inevitable, EU companies pursued win-win strategies by seeking first mover advantages. In this case industry did not receive substantial direct aid from national governments or the EU, but as part of a long-term strategic approach such measures might feature more prominently in future episodes.

Risks of this approach

Does the strategic environmental policy approach really provide an adequate foundation upon which to overcome the competitiveness problem and construct the next generation of environmental policies? Or does reliance on win-win solutions offer more of a false panacea than a source of legitimation?[8] Unfortunately, a combination of economic considerations, legal impediments and moral uncertainties make it unfeasible for the Commission or ambitious member states to justify tougher environmental standards solely on the grounds of positive competitive gains.

As discussed above, the fact that studies of how environmental regulation affects competitiveness are inconclusive, and that negative economic impacts on certain sectors are hard to deny, highlights the unlikelihood of obtaining win-win solutions in each and every case (Stewart 1993: 2082–2084, Jaffe *et al.* 1995: 157, Palmer *et al.* 1995, EEA 1996: 33). Evidence suggests that, if they occur at all, the economic gains from stringent environmental regulation often fail to offset the costs completely. Moreover, the negative relationship may intensify as global infrastructure improves and competitive advantage depends more on small factors such as the marginal costs of environmental protection (Medhurst 1993: 43).

And even if one assumes that in many cases a strategic environmental approach could produce competitive advantages, important economic problems of distribution arise because win-win offsets do not always involve gains by the industry being regulated. Rather, the winners and losers under a win-win approach are often different groups (EEA 1996: 33–34). For example, the Commission might be correct that tough unilateral environmental standards will induce first mover advantages and help the EU environmental protection industry capture the lucrative market for green technology and services, but this 'win' will be accompanied by losses imposed on dirty and often declining industries which must purchase this technology and pay higher costs for energy, waste disposal or emissions reductions, which could force them to relocate or close. Devising and coordinating a system of 'strategic' environmental regulation which balances winners and losers therefore involves much larger questions about the extent to which governments should involve themselves in market intervention and industrial policy.

Of course, capturing green technology markets is only one element of win-win solutions – ideally, stringent environmental standards encourage efficient production and induce innovative management, the savings from which accrue to the regulated industry itself. Moreover, a firm's capacity for technological innovation is increasingly considered the fundamental determinant of its competitiveness. Nevertheless, difficult questions of distribution and industrial policy remain because these offsets are often contingent upon targeted government tax rebates and subsidies, as shown particularly clearly in the case of carbon–energy taxes (Chapter 5) (see also EC 1996b, Golub 1998).

Coordinated packages of direct offsets might increase the frequency of win-win solutions, but strategic environmental policies which incorporate these features encounter several legal impediments. The Polluter Pays Principle (PPP), supported by the OECD and enshrined in Article 130 of the Maastricht Treaty, militates against government, as opposed to industry, investment in pollution abatement. GATT rules on subsidies and countervailing measures allow a certain amount of 'assistance to promote adaptation of existing facilities to new environmental requirements imposed by law', but 'cannot cover investments that result in manufacturing cost savings' and generally discourage any green subsidies which are open to abuse and threaten to undermine the PPP (Charnovitz 1993, Esty 1994: 170, EC 1996a: 15). GATT rules also take a dim view of achieving win-win solutions through green protectionism, and EU actions to develop ecolabels and to protect animals and forests have received international criticism precisely because they privilege EU industry. In sharp contrast to these cases, it should be noted, changes to GATT rules as part of ongoing trade liberalisation actually facilitated the greening of the CAP (Chapter 7).

Direct government intervention, whether in the form of industrial policy, recycled tax revenue or targeted subsidies, also runs a high risk of violating

EU treaty provisions on competition and state aids (EC 1996a, 1996b, Delbeke and Bergman 1998).[9] Even if used to facilitate win-win solutions, the Commission recognises that 'state aid is liable to give certain firms an advantage over their competitors in other member states not receiving such aid' (EC 1995b: 156). EC rules, while highly complex and open to various interpretations, generally allow governments to cover only 15–25 per cent of a firm's investment towards meeting new environmental standards.

Ironically, even if the strategic approach were economically feasible and permissible under international trade law, its premise actually undermines at least part of its legal legitimacy under EU treaty provisions (Golub 1996a). One of the traditional justifications for harmonising EU environmental laws, and under the subsidiarity principle one necessary to establish that EU action is 'better' than national action, has been that diverging national regulations convey illegitimate competitive advantages to laggard states (often in southern Europe) by creating room for environmental dumping and pollution havens. This argument from economic considerations is no longer tenable if, as the Commission now contends, stringent unilateral standards by progressive member states actually provide them with competitive advantages. The only remaining economic rationale for regulation at the EU level would be to prevent green protectionism amongst member states – the leaders rather than the laggards would constitute the threat to competition. While this could legitimate EU product standards, it would not extend legitimation to EU regulation of industrial processes.

Moreover, with limited internal competence for certain types of environmental regulation, the Commission will have a more difficult time establishing the legal legitimacy of an international EU voice (Nollkaemper 1987). The Commission's dilemma is only made worse by the fact that several aspects of EU win-win solutions do not require any international action. Continued growth in the market for green technology requires that other states eventually adopt tough environmental laws similar to those in the EU, which would justify a strong Commission role in international negotiations, but the economic efficiency and other cost savings associated with win-win solutions can be captured simply with national or possibly EU rules.

Finally, and perhaps most significantly, the strategic, 'free lunch', approach poses a moral question: does the Commission really want to stake the legitimacy of future EU environmental policy primarily on the availability of win-win solutions? To do so would significantly curtail the range of permissible measures, and would surrender the classic source of legitimation for regulation designed to safeguard public goods – that it provides net social benefits. This is not a win-win approach, but rather a 'win-loss' perspective, which would justify stringent regulation when the social welfare it provides through a clean environment outweighs the social cost, including a possible loss of economic competitiveness (compare Palmer *et al.* 1995 with Porter and van der Linde 1995; see EC 1996b, Golub 1998).[10]

EU competitiveness and the environment: the difficult road ahead

Twenty-five years of internal EC regulation and international agreements have done much to improve environmental protection. How has this been accomplished given the inherent tension with economic objectives? Green leadership from northern European states, often in alliance with parts of the Commission and usually with the persistent support of the European Parliament's environment committee, helps explain many such policy advances (Judge 1993, Héritier *et al*. 1996, Sbragia 1996b).

However, the chapters in this volume highlight two recurring themes which environmental groups often lament, and which provide perspective on the achievements made to date as well as the path ahead: that competitiveness objectives have delayed and in some cases prevented environmental regulation at the national level, and that reconciling these conflicting objectives through collective international action has proven both politically and legally difficult. Judging from the seven cases here, as well as those considered in the other two volumes of this series, this task will only become more difficult in the coming years.

Constructing the next stage of EU environmental policy appears daunting in the face of pressures to maintain and improve economic competitiveness, but atrophy of supranational authority is not necessarily inevitable, nor environmental decline unavoidable. Continuation of a strong EU role in this field depends on convincing a majority of states that supranational action remains the most rational means of combating environmental degradation, and that economic competitiveness can in fact be reconciled with environmental protection. To rebuild confidence in the European project, the Commission has embraced two strategies.

The first is to minimise regulatory costs by replacing the traditional EU command and control approach with a greater emphasis on new flexible and efficient policy tools such as ecotaxes, ecolabels and voluntary agreements with industry (Collier 1997, Golub 1997, 1998). The second has been to transform the competitiveness debate in favour of EU environmental action by seeking new sources of legitimacy for environmental regulation. In the cases where they can be found, and legally achieved, win-win solutions should be seized as a means of reconciling competitiveness and environmental protection. But free lunches may constitute a minority of cases, particularly if EU and GATT rules prevent the possibility of wedding environmental and industrial policy. Tough unilateral EU environmental standards therefore run the very real risk of damaging EU economic competitiveness. One possible way forward, then, would be for the Commission to strengthen its commitment to multilateral solutions, not only because they 'offer the most effective way to tackle global and transboundary environmental problems' (EC 1996a: 10–11), but also because, as examples of the 'stick' approach, they offer an attractive means of saving the global commons without sacrificing competitiveness.

Even the multilateral approach has it limits, however, some of which are identified by the authors in this volume. First, the Commission will continue to encounter resistance over the issue of its external competence. In shipping (Chapter 4), member states have traditionally resisted empowering the Commission as an external voice in the IMO, preferring wide latitude for their individual negotiating efforts, although Nollkaemper identifies some recent moves towards greater Commission involvement. Pallemaerts finds similar instances in the early stages of pesticide regulation (Chapter 3) where the Council denied the Commission's request to play a leading role in international negotiations. Similarly, member states are not convinced of the value-added from EU environmental development assistance, and outside of a few exceptional cases jealously preserve traditional bilateral aid arrangements, with their attendant advantages for pursuing national security interests and promoting national exports. Second, multilateral solutions also pose sometimes insurmountable collective action problems, which is why Heller advocates a 'minilateral' solution to CO_2 reduction through the use of JI (Chapter 5).

A third problem is that while international harmonisation and other multilateral arrangements are often a solution to avoid competitive disadvantages, they are sometimes a second-best option because they undermine the stringency of EU laws. In essence, rather than wielding a 'stick' at the international level, the EU is forced to capitulate on environmental standards. For instance, international adoption of the ISO standard for ecolabels has been suggested as a way to avoid a proliferation of EU rules which constitute trade barriers, but its environmental merits have been seriously questioned in comparison to the EU's preferred labelling criteria (Europe Environment 1995, Taschner 1998). Creating ISO 'humane standards' which, when met by hunters, would guarantee the export of their furs, might obviate recourse to trade restrictions from an outright EU ban on the use of leg-hold traps, but might not adequately prevent the use of these traps in North America. Finally, the Commission's sanguine view that multinational agreements will be deemed trade-friendly under GATT/WTO rules (EC 1996a: 17) may not materialise in practice, in which case resolving competitiveness issues will require a different solution. This raises interesting questions about what the EU should do in such situations, whether it can maintain its own high standards, whether it must adjust downwards in order to continue trade relations with its partners, and whether at some point it should abandon trade for environmental reasons.

In light of all these concerns, must proponents of a cleaner environment therefore concede defeat? Not necessarily. As mentioned earlier, the win-win and 'stick' approaches do not exhaust the possible grounds upon which to legitimate aggressive national and EU environmental regulation. Despite its zealous commitment to free lunches as a means of reconciling economic and environmental objectives, the Commission has been careful not to abandon entirely the possibility of justifying pollution reduction in the absence of similar foreign measures, even without positive effects on competition. The fifth EAP includes a section on costing environmental pollution (and the social costs of 'non-environment') which espouses a more traditional philosophy: the only condition that is really required to legitimate environmental protection is that 'the ultimate benefits should outweigh the so-called costs over time' (EC 1993b: 141). This demands consideration for the welfare of future generations as well as the Treaty's precautionary principle. Meeting the 'ultimate benefits' test should not prove difficult given the enormous social costs of forgoing environmental action (EP 1994).[11]

Pursuing environmental protection for its intrinsic value has long been the primary motivation of leader states in the EU as well as the Commission, and finds clear expression in both the Single European Act and the Maastricht Treaty on European Union. It has underpinned previous policy developments, and it might offer the best rationale for combating a wide range of polluting activities in the future. And, as Nick Robins argues, the adoption of a traditional rather than a strategic approach to policymaking is also crucial for the continuation of environmental and humanitarian aid. Justifying member state and EU assistance strictly on the grounds of potential win-win solutions not only abdicates moral leadership and abandons aid efforts whose primary objective is poverty relief, it also commits donor states to a system which yields perverse results such as those evident in the Pergau Dam incident. This perspective diverges radically from Heller's plea to substitute a 'culture of business for a culture of aid' (Chapter 5).

But before deciding to pursue a unilateral approach, regardless of possible adverse effects on competitiveness, the Commission and individual member states must also consider which course of action will actually maximise environmental protection. If firms respond to declining competitiveness by relocating or reinvesting outside the EU, the important question becomes whether such industrial 'flight' degrades or improves the global environment. The former suggests that strict regulation within the EU will simply result in greater damage abroad as pollution shifts location (for several examples, including the US–Mexico case, see French 1993). Alternatively, flight might not be a bad thing for the environment, as multinational companies fleeing proposed legislation in one country often take their state-of-the-art methods and technology with them. Some studies have found this 'environmental technology transfer' effect whether examining investment in Latin America, East Asia, Eastern Europe or the former Soviet Union (Stewart 1993: 2070, Zamparutti and Klavens 1993: 124, Jaffe et al. 1995, Palmer et al.1995: 130). Additional research is required to illuminate this facet of the environment–competitiveness issue.

The path forward is most certainly paved with obstacles. Inevitably, despite the fact that excessive attention to competitiveness may be myopic and misguided, appeals to general social welfare will be more difficult, even impossible, in a political climate sensitised to any legislation which might undermine comparative advantage. At the same time, however, neither a strategic nor a multilateral approach by itself offers a satisfactory solution. Those devising the next generation of environmental policies in the EU therefore face the difficult challenge of securing political viability for a regulatory package which strikes an appropriate balance between competitiveness and environmental protection.

Notes

1 The literature devoted to international environmental policy, when it addresses the role of the EU, its member states, and competitiveness, does so as a means to explore regime building, interest group involvement and the environmental implications of individual international conventions (Young 1989, Hurrell and Kingsbury 1992, Haas, Keohane and Levy 1993). Moreover, the leading texts on the determinants and future of European economic competitiveness pay little or no attention to the issue of environmental policy (Francis and Tharakan 1989, Hughes 1993). For analyses which do discuss competitiveness issues as they relate to both internal and external EC environmental policy, see Freeman et al. (1991) and Klepper (1992).
2 Precisely how one should conceptualise competitiveness, particularly its relationship with productivity, remains a matter of intense debate, as does the notion that a country's economic competitiveness depends on the performance of firms within its boundaries (Francis 1989, Hughes 1993, Krugman 1994).
3 Heerings attributes the decline of the EC fertiliser industry in part to mounting pressure, particularly in Belgium and the Netherlands, to reduce cadmium and phosphorus discharges. The example highlights the role of national regulation, as the 1983 EC Directive on cadmium emissions explicitly exempted the phosphate fertiliser industry.
4 Heller's 'window of opportunity' prognosis sharply contradicts scholars who suggest that action can be postponed another decade or more without risking environmental catastrophe (see Beckerman 1991).
5 In its 1992 'Communication on Industrial Competitiveness and Environmental Protection', the Commission claimed that it 'fully support[ed] the basic rule according to which a country should not unilaterally restrict imports on the basis of environmental damage which does not impact on a country's territory' (EC 1992: 8).
6 To cite just one of the many examples from Commission documents, the Fifth EAP contains a section on international competitiveness which portrays stringent environmental requirements as sources of competitive advantage (EC 1993b: 64). In 1990 the world market for pollution control was $200 billion. The EC market has been estimated at $50 billion, and enormous growth is forecast through year 2000. Commission figures also show that 600,000 European jobs are linked to the eco-industry (EC 1992).
7 Robins also identifies conditions under which development aid promotes rather than harms recipient states, such as where national governments have used aid finance to raise the competitiveness of developing country producers, assisting

them to improve production standards and achieve certification so as to gain access to the environmentally conscious markets of the North.

8 This charge has been levelled against Martin Porter and other American proponents of this approach (Oates *et al.* 1993, Palmer *et al.* 1995).

9 Articles 92–94 of the Treaty prohibit state aid which distorts competition within the Common Market.

10 The distinction between this approach and the win-win strategy is crucial: 'the suggestion of proponents of the Porter hypothesis is *not* that the benefits of environmental regulation (in terms of reduced health and ecological damages) exceed the costs of environmental protection. . . . Rather, the notion of a 'free lunch' is that – putting aside the benefits of environmental protection – the costs of regulatory action can be zero or even negative (a 'paid lunch')' (Jaffe *et al.* 1995:155).

11 For a quantitative study of US air and water laws which finds that social gains from environmental regulation have outweighed the costs, see Burtraw and Portney (1991).

Bibliography

Arden-Clarke, Charles (1993) 'Environment, Competitiveness and Countervailing Measures,' in OECD, *Environmental Policies and Industrial Competitiveness* (Paris: OECD).

Barbera, Anthony and Virginia McConnell (1990) 'The Impact of Environmental Regulations on Industry Productivity: Direct and Indirect Effects,' *Journal of Environmental Economics and Management*, Vol. 18, No. 1, pp. 50–65.

Barrett, Scott (1993) 'Strategic Environmental Policy and International Competitiveness,' in OECD, *Environmental Policies and Industrial Competitiveness* (Paris: OECD).

Bartik, Timothy (1988) 'The Effects of Environmental Regulation on Business Location in the United States,' *Growth Change*, Vol. 19, No. 3, pp. 22–44.

Beckerman, Wilfred (1991) 'Global Warming: A Sceptical Economic Assessment,' in D. Helm (ed.), *Economic Policy Towards the Environment* (Oxford: Blackwell).

Blazejczak, Jürgen (1993) 'Environmental Policies and Foreign Investment: The Case of Germany,' in OECD, *Environmental Policies and Industrial Competitiveness* (Paris: OECD).

Burtraw, Dallas and Paul Portney (1991) 'Environmental Policy in the United States,' in D. Helm (ed.), *Economic Policy Towards the Environment* (Oxford: Blackwell).

Charnovitz, Steve (1993) 'Environmental Trade Measures and Economic Competitiveness: An Overview of the Issues,' in *Environmental Policies and Industrial Competitiveness* (Paris: OECD).

——— (1994) 'Free Trade, Fair Trade, Green Trade: Defogging the Debate,' *Cornell International Law Journal*, Vol. 27, pp. 460–525.

Choucri, Nazli (1993) 'Multinational Corporations and the Global Environment,' in Nazli Choucri (ed.), *Global Accord: Environmental Challenges and International Responses* (Cambridge, MA: MIT Press).

Collier, Ute (ed.) (1997) *Deregulation in the European Union: Environmental Perspectives* (London: Routledge).

Collier, Ute and Jonathan Golub (1997) 'Environmental Policy and Politics,' in Martin Rhodes, Paul Heywood and Vincent Wright (eds), *Developments in West European Politics* (Basingstoke: Macmillan).

Conrad, Klaus and Dieter Wastl (1995) 'The Impact of Environmental Regulation on Productivity of German Industries,' Mattei Foundation Working Paper 9/95, (Milan: Fondazione ENI Enrico Mattei).

Cremer, Wolfgang and Andreas Fisahn (1998) 'New Environmental Policy Instruments in Germany,' in Jonathan Golub (ed.), *New Instruments for Environmental Protection in the EU* (London: Routledge).

Dean, Judith (1992) 'Trade and the Environment: A Survey of the Literature,' in Patrick Low (ed.), *International Trade and the Environment* (Washington, DC: International Bank for Reconstruction and Development/World Bank).

Delbeke, Jos and Hans Bergman (1998) 'Environmental Taxes and Charges in the EU,' in J. Golub (ed.), *New Instruments for Environmental Protection in the EU* (London: Routledge).

DRI (1994) *Potential Benefits of Integration of Environmental and Economic Policies* (London: Graham and Trotman).

EC (1982) 'The Competitiveness of European Community Industry,' Brussels, 5 March.

—— (1987) *Improving Competitiveness and Industrial Structures in the Community* (Luxembourg: Office for Official Publications of the European Communities).

—— (1992) 'Communication on Industrial Competitiveness and Environmental Protection,' SEC(92)1986, 24 November (Brussels: Commission of the European Communities).

—— (1993a) 'White Paper on Growth, Competitiveness and Employment', *Bulletin of the European Communities*, supplement, No. 6/93.

—— (1993b) 'Towards Sustainability,' Fifth Environmental Action Programme (Luxembourg: Commission of the European Communities).

—— (1994) 'Communication on Economic Growth and the Environment: Some Implications for Economic Policymaking,' COM(94)465, 3 November (Brussels: Commission of the European Communities).

—— (1995a) 'Report of the Group of Independent Experts on Legislative and Administrative Simplification,' COM(95)288, 21 June (Brussels: Commission of the European Communities).

—— (1995b) *Competition Law in the European Communities Volume IIA, Rules Applicable to State Aid* (Brussels: Commission of the European Communities).

—— (1996a) 'Communication on Trade and the Environment,' COM(96)54, 28 February (Brussels: Commission of the European Communities).

—— (1996b) 'Economic Incentives and Disincentives for Environmental Protection,' Proceedings of European Commission and Council Presidency Conference, Rome, 7 June.

EEA (1996) *Environmental Taxes. Implementation and Environmental Effectiveness* (Copenhagen: European Environmental Agency).

EP (1993a) 'Report on Environment and Trade,' 22 January (Brussels: European Parliament).

—— (1993b) 'Resolution on Environment and Trade,' A3–0329/92 (Brussels: European Parliament).

—— (1994) 'Report of the Committee on the Environment, Public Health and Consumer Protection on the Need to Assess the True Costs to the Community of "Non-Environment",' *European Parliament Document A3–0112/94*, 23 February.

Esty, Daniel (1994) *Greening the GATT* (Washington DC: Institute for International Economics).

Europe Environment (1995) 'EEB Mounts Campaign Against ISO Standards,' *Europe Environment*, No. 465, 14 November (Brussels: Europe Information Service).

Francis, Arthur (1989) 'The Concept of Competitiveness,' in A. Francis and P. K. M. Tharakan (eds), *The Competitiveness of European Industry* (London: Routledge).

Francis, Arthur and P.K.M. Tharakar (eds) (1989) *The Competitiveness of European Industry* (London: Routledge).

Freeman, Christopher, Margaret Sharp and William Walker (eds) (1991) *Technology and the Future of Europe: Global Competition and the Environment in the 1990s* (London: Pinter).

French, Hilary (1993) 'Costly Tradeoffs: Reconciling Trade and the Environment,' Worldwatch Paper 113 (Washington, DC: Worldwatch Institute).

Golub, Jonathan (1996a) 'Sovereignty and Subsidiarity in EU Environmental Policy,' *Political Studies*, Vol. 44, No. 4, pp. 686–703.

—— (1996b) 'British Sovereignty and the Development of EC Environmental Policy,' *Environmental Politics*, Vol. 5, No. 3, pp. 700–728.

—— (1996c) 'Why Did They Sign? Explaining EC Environmental Policy Bargaining,' RSC Working Paper No. 96/52 (Florence: European University Institute).

—— (1997) 'Recasting EU Environmental Policy: Subsidiarity and National Sovereignty,' in U. Collier, J. Golub and A. Kreher (eds), *Subsidiarity and Shared Responsibility: New Challenges for EU Environmental Policy* (Baden-Baden: Nomos).

—— (ed.) (1998) *New Instruments for Environmental Protection in the EU* (London: Routledge).

Grossman, Gene and Alan Kreuger (1993) 'Environmental Impacts of a North American Free Trade Agreement,' in Peter Garber (ed.), *The US–Mexico Free Trade Agreement* (Cambridge, MA: MIT Press).

Haas, Peter, Robert Keohane and Marc Levy (eds) (1993) *Institutions for the Earth: Sources of Effective International Environmental Protection* (Cambridge, MA: MIT Press).

Haigh, Nigel (1984) *EEC Environmental Policy and Britain: an Essay and a Handbook* (London: Environmental Data Services).

—— (1989) *EEC Environmental Policy and Britain*, second edn (London: Institute for European Environmental Policy).

Heerings, Hans (1993) 'The Role of Environmental Policies in Influencing Patterns of Investments of Transnational Corporations: Case Study of the Phosphate Fertiliser Industry,' in OECD, *Environmental Policies and Industrial Competitiveness* (Paris: OECD).

Héritier, Adrienne, Christoph Knill and Susanne Mingers (1996) *Ringing the Changes in Europe: Regulatory Competition and the Transformation of the State: Britain, France, Germany* (Berlin: de Gruyter).

Hildebrand, Philipp (1993) 'The European Community's Environmental Policy, 1957 to 1992: From Incidental Measures to an International Regime?', in David Judge (ed.), *A Green Dimension for the European Community: Political Issues and Processes* (London: Frank Cass).

Hughes, Kirsty (ed.) (1993) *European Competitiveness* (Cambridge: Cambridge University Press).

Hurrell, Andrew and Benedict Kingsbury (eds) (1992) *The International Politics of the Environment: Actors, Interests and Institutions* (Oxford: Clarendon Press).

Jaffe, Adam, Steven Peterson, Paul Portney and Robert Stavins (1995) 'Environmental Regulation and the Competitiveness of US Manufacturing: What Does the Evidence Tell Us?', *Journal of Economic Literature*, Vol. 33, pp. 132–163.

Judge, D. (1993) 'Predestined to Save the Earth: The Environment Committee of the European Parliament,' in D. Judge (ed.), *A Green Dimension for the European Community: Political Issues and Processes* (London: Frank Cass).

Klepper, Gernot (1992) 'The Political Economy of Trade and the Environment in Western Europe,' in Patrick Low (ed.), *International Trade and the Environment* (Washington, DC: The World Bank), pp. 247–259.

Krugman, Paul (1994) 'Competitiveness: A Dangerous Obsession,' *Foreign Affairs*, Vol. 73, No. 2, pp. 28–44.

Leonard, Jeffrey (1988) *Pollution and the Struggle for the World Product* (Cambridge: Cambridge University Press).

McConnell, Virginia and Robert Schwab (1990) 'The Impact of Environmental Regulation on Industry Location Decisions: The Motor Vehicle Industry,' *Land Economics*, Vol. 66, No. 1, pp. 67–81.

Morris, Robert (1993) 'A Business Perspective on the Competitiveness Effects of Environmental Policies,' in OECD, *Environmental Policies and Industrial Competitiveness* (Paris: OECD).

Nishijima, Yoichi (1993) 'Internalisation of Environmental Costs and Competitiveness in Japan,' in OECD, *Environmental Policies and Industrial Competitiveness* (Paris: OECD).

Nollkaemper, André (1987) 'The European Community and International Environmental Cooperation – Legal Aspects of External Community Powers,' *Legal Issues of European Integration*, pp. 55–91.

Oates, Wallace *et al.* (1993) 'Environmental Regulation and Environmental Competitiveness: Thinking About the Porter Hypothesis,' Resources for the Future Discussion Paper No. 94–02 (Washington, DC: Resources for the Future).

OECD (1996) *Industrial Competitiveness* (Paris: OECD).

Palmer, Karen, Wallace Oates and Paul Portney (1995) 'Tightening Environmental Standards: The Benefit-Cost or the No-Cost Paradigm?', *Journal of Economic Perspectives*, Vol. 9, pp. 119–132.

Pearson, Charles (1987) *Multinational Corporations, Environment and the Third World* (Durham, NC: Duke University Press and World Resources Institute).

Porter, Michael (1990) *The Competitive Advantage of Nations* (New York: Free Press).

Porter, Michael and Claas van der Linde (1995) 'Toward a New Conception of the Environment–Competitiveness Relationship,' *Journal of Economic Perspectives*, Vol. 9, pp. 97–118.

Rehbinder, Eckard and Richard Stewart (1985) *Environmental Policy Protection, European University Institute Integration Through Law Series* (Berlin: de Gruyter).

Richardson, D. and C. Rootes (1995) *The Green Challenge: the Development of Green Parties in Europe* (London: Routledge).

Robins, Nick (1996) 'Greening European Foreign Policy: From Environmental Insecurity to Diplomacy for Sustainable Development,' paper presented at The Green Agenda for the IGC Conference, University Association for Contemporary European Studies, London, 29 March.

Rudnick, David (1992) 'Taken to the Cleaners,' *Financial Times*, 16 June, pp. 26–27.

St.-Pierre, Antoine (1993) 'Industrial Competitiveness, Trade and the Environment in Canada,' in OECD, *Environmental Policies and Industrial Competitiveness* (Paris: OECD).

Sbragia, Alberta (1996a) 'Institution Building from Above and Below: The European Community in Global Environmental Politics,' mimeo.

—— (1996b) 'The Push–Pull of Environmental Policymaking,' in H. Wallace and W. Wallace (eds), *Policymaking in the European Union*, third edn (Oxford: OUP).

Stewart, Richard (1993) 'Environmental Regulation and International Competitiveness,' *Yale Law Journal*, Vol. 102, No. 8, pp. 2039–2106.

Taschner, Karola (1998) 'Environmental Management and Audit Systems: The European Regulation,' in Jonathan Golub (ed.), *New Instruments for Environmental Protection in the EU* (London: Routledge).

Van der Linde, Claas (1993) 'Micro-Economic Implications of Environmental Regulations: A Preliminary Framework,' in OECD, *Environmental Policies and Industrial Competitiveness* (Paris: OECD).

Verbruggen, Harmen (1993) 'The Trade Effects of Economic Instruments,' in OECD, *Environmental Policies and Industrial Competitiveness* (Paris: OECD).

Vogel, David (1993) 'The Making of EC Environmental Policy,' in S. Anderson and K. Eliassen (eds), *Making Policy in Europe* (London: Sage).

Young, Oran (1989) *International Cooperation: Building Regimes for Natural Resources and the Environment* (Ithaca, NY: Cornell University Press).

Zamparutti, Anthony and Jon Klavens (1993) 'Environment and Foreign Investment in Central and Eastern Europe: Results from a survey of Western Corporations,' in OECD, *Environmental Policies and Industrial Competitiveness* (Paris: OECD).

2

EU POLICY FOR OZONE LAYER PROTECTION

Ian Rowlands

Debates about 'sustainable development' during the past decade have high-lighted the intimate interconnection between 'environment' and 'economy'.[1] It is now widely agreed that concerns about one cannot be divorced from concerns about the other. Environmental policy obviously has profound implications for the economy; just as economic policy has ramifications for the quality of the environment, and efforts to improve the same.[2]

Given the greater acceptance of these links, analysts have paid more atten-tion during the past ten years to the ways in which the two interact. This chapter aims to contribute to this general debate by looking at one particular question: to what extent have concerns about economic competitiveness, and the interests of industry more generally, influenced the development of the European Union's policy on ozone layer depletion?

By focusing upon this rather specific question, it is anticipated that a number of different areas will be opened up for exploration. Most obviously, the concerns of the Union's producers and users of ozone-depleting substances will be revealed – not only the factors that have affected percep-tion of their own interests, but also the ways in which they have tried to exert control over policy reform. The question also directs attention to the decision-makers – that is, it invites us to investigate how decision-makers have reacted to these pressures.

The chapter is divided into five subsequent sections. The following section presents a brief overview of the issue of ozone layer depletion. It focuses, in particular, upon the EC/EU response, while nevertheless placing it within the broader global context. The next three sections analyse a number of individual debates, broadly following a chronological order. More specifically, they focus upon the regulation of chlorofluorocarbons, hydrochlorofluorocarbons and a few more recent concerns, which include the regulation of methyl bromide. The chapter concludes with a summary, as well as some tentative answers to the questions initially posed.

34

Overview

During the past quarter of a century, the issue of ozone layer depletion has attracted considerable scientific and political attention. Though not without its ups and downs, the international community has steadily strived to reach progressively more restrictive agreements on the production and use of the chemicals that serve to deplete the planet's protective layer of stratospheric ozone. A full examination of the issue's development is, of course, beyond the scope of this chapter (Morrisette 1989, Parson 1993, Parson and Greene 1995, Rowlands 1995a). What is useful for the purposes of this particular study, however, is a review of the key decisions emerging from both global and European fora.

Table 2.1 lists the most significant global agreements. Of particular note in this list is the landmark 1987 Montreal Protocol, along with its subsequent amendments. Table 2.2, meanwhile, lists the outputs arising from European discussions. Since 1988, they have mainly taken the form of Regulations, as European decision-makers have decided how the global agreements should be implemented, and possibly built upon.[3]

As the purpose of this section is to lay the backdrop for the rest of the chapter, it is also useful to remind ourselves of the 'balance of power' between the Commission and the member states throughout the history of the ozone layer depletion issue. This is crucial to ensure a full understanding of the development of the European response.

When international debate about the ozone layer began – during the mid-1970s – 'it was not clear that the EC was competent to negotiate in these matters' (Jachtenfuchs 1990: 262). This, however, did not prevent the Commission from participating in the ongoing international negotiations, and, indeed, signing the Vienna Convention when it was opened for signature in March 1985.[4] Of most significance, however, was a decision taken at a meeting of the EC Environment Council on 24 November 1986. At that time, the member states authorised the Commission to take part, on behalf of the Community, in the negotiations towards a protocol. The Commission was, however, presented with a well-defined mandate (Jachtenfuchs 1990: 265) – a mandate which, under the terms of the Treaty of Rome, had to be agreed unanimously by the Council.

This has continued to be the general trend since that time – that is, the Council of Ministers has debated a particular position, which, when agreed, has been presented to the Commission. The Commission has, in turn, negotiated on behalf of the Community in the international meetings. When agreements have been reached at this level – at so-called 'Meetings of the Parties to the Montreal Protocol' – the issue has returned to the Council of Ministers. The member states' environment ministers have had to decide how the commitments agreed at the international forum should be translated into EC and member state law (and whether the Community should go

Table 2.1 Major international agreements

International agreement	Key dates and status	Major commitments
Vienna Convention on Depletion of the Ozone Layer	• opened for signature: 22 March 1985 • entered into force: September 1988 • current status: 165 ratifications (at 25 August 1997)	none
Montreal Protocol on Substances that Deplete the Ozone Layer	• opened for signature: 16 September 1987 • entered into force: January 1989 • current status: 162 ratifications (at 25 August 1997)	• freezing of consumption of five CFCs by 1992 and 50 per cent reduction by 1999 • freezing of consumption of three halons by 1992
London Adjustments and Amendments to the Montreal Protocol	• agreed: 29 June 1990 • amendments entered into force: August 1992 • current status: 116 ratifications (at 25 August 1997	• elimination of 15 CFCs, three halons and carbon tetrachloride by 2000 • elimination of methyl chloroform by 2005
Copenhagen Adjustments and Amendments to the Montreal Protocol	• agreed: 25 November 1992 • amendments entered into force: June 1994 • current status: 72 ratifications (at 25 August 1997)	• elimination of 15 CFCs, carbon tetrachloride, methyl chloroform and HBCs by 1996 • elimination of three halons by 1994 • freeze of HCFCs by 1996, with their eventual elimination by 2030 • freeze of methyl bromide by 1995
Vienna Adjustments to the Montreal Protocol	• agreed: 7 December 1995	• lower cap on HCFCs, but elimination date remains at 2030 • elimination of methyl bromide by 2010, with 25 per cent cut by 2001 and 50 per cent cut by 2005

Sources: The Vienna Convention, the Montreal Protocol, the London Adjustments and Amendments, and the Copenhagen Adjustments and Amendments are reprinted in *International Legal Materials*. Respectively, each appears in Vol. 26 (1989), pp. 1516–1540; Vol. 26 (1989), pp. 1541–1561; Vol. 30 (1991), pp. 537–554; and Vol. 32 (1993), pp. 874–887. The Vienna Adjustments may be found on the World Wide Web.

Table 2.2 Major European agreements

European agreement	Date agreed	Major commitments
Resolution (C133)	30 May 1978	• calls for industry not to increase production capacity of two CFCs
Decision 80/372	26 March 1980 (OJ L90, 3 April 1980)	• calls on all Member States to take all appropriate measures to ensure that there is no increase in production capacity of two CFCs • Member States are to achieve a 30 per cent reduction in the use of CFCs in aerosol cans by 1982
Decision 82/795	16 November 1982 (OJ L329, 25 November 1982)	• repeats obligation of Decision 80/372 • adds a definition of production capacity and a reference figure for annual Community CFC production capacity (480,000 tonnes)
Decision 88/540 and Regulation 3322/88	14 October 1988 (OJ L297, 31 October 1988)	• enables the Community to ratify the Vienna Convention and the Montreal Protocol • implements the terms of the Montreal Protocol
Regulation 594/91	4 March 1991 (OJ L67, 14 March 1991)	• elimination of CFCs by mid-1997, carbon tetrachloride by 1998, halons by 2000 and methyl chloroform by 2006
Regulation 3952/92	30 December 1992 (OJ L405, 31 December 1992)	• elimination of halons by 1994, CFCs and carbon tetrachloride by 1995 and methyl chloroform by 1996
Regulation 3093/94	15 December 1994 (OJ L333, 22 December 1994)	• freezing of methyl bromide production by 1995, and 25 per cent reduction by 1998 • elimination of HCFC consumption by 2015, with five intermediate targets (cap set at 2.6 per cent of CFC consumption and HCFC consumption in 1989) • elimination of HBCs by 1996

beyond the terms of the just-reached international commitment). A development worthy of note, however, is that the Single European Act of 1987 introduced the possibility of qualified majority voting; this has in fact been used in the development of policy on ozone layer depletion. Moreover, the Treaty on European Union of 1992 provided for even further input – stronger powers of veto and amendment – by the European Parliament

(Chance 1995); though to a limited extent, Parliament has nevertheless also been involved in the ozone layer depletion issue.

Controlling CFCs

Discussions about a potential 'can ban', 1974–1985

When significant concern about potential ozone layer depletion first arose, chlorofluorocarbons (CFCs) were the set of chemicals initially deemed most culpable.[5] The United States was the first country to react. In 1977, a 'can ban' was instituted there – that is, a restriction upon the use of CFCs as propellants in non-essential aerosol sprays. Soon afterwards, US representatives began to push for a similar global ban.[6] With this, we see the first connection between the European Community's external policy on the ozone layer and its concern for competitiveness at a variety of levels. (Table 2.3 reveals that Community production of CFCs in 1974 accounted for one-third of the world total, though somewhat less than that of the United States.)

The Community response at this time was cautious, to say the least. As outlined in Table 2.2, Community agreements fell short of the desired can ban; indeed, the production cap adopted in 1980 had little impact (because European industry was operating well below capacity at this time), and the 30 per cent reduction in the use of CFCs in aerosol cans was easily achievable because use figures were already declining (in response to changes that had already occurred). In fact, Haigh suggests that 'there is reason to believe that the figure of 30 per cent was chosen because it was known that it could be achieved without creating too much difficulty for industry' (Haigh 1989: 268).

Can EC prudence during the late 1970s really be explained by concerns for industrial competitiveness? Two immediate reasons – suggesting that any can ban in the EC would have been much more economically damaging than it had been in the USA – suggest that it can. First, more CFC

Table 2.3 Global production of CFC-11 and CFC-12 by country/region, 1974

Country/Region	Percentage share
United States	44
European Communities	33
Japan	8
USSR and Eastern Europe	8
Other OECD	5
Developing Countries	2

Sources: Compiled from OECD (1976), Gamlen *et al.* (1986) and Thornton (1990)
Note: All calculations by weight

production went to the aerosol spray industry in the EC as compared to the US – in both relative and absolute terms.[7] As a consequence, an EC ban upon the use of CFCs in non-essential aerosols would affect a greater proportion of its industry, and could thereby be more economically disruptive in Europe than it had been in the United States.

And second, the international trade in CFCs was very lucrative for a number of European states during the 1970s. Indeed, compared with the US, EC countries exported a higher share of their CFC production. The value of British exports, for example, was placed at [UK] £70 million in 1974 (DoE 1976: 6). Thus, if the EC states were to restrict production, then substantial export markets might have to be forsaken. Explicitly citing this as a rationale for its own *laissez-faire* policy on CFCs, the UK Minister of State for the Environment, Denis Howell, said in 1975 that an aerosol ban could cause 'a considerable loss to our balance of payments' and have 'far-reaching repercussions on the aerosol industry' (NS 1975: 336).

More generally, reports from both national governments and the Commission itself highlighted the detrimental economic impact that further regulation could have. As an example of the former, the British Department of the Environment estimated that 50,000 jobs were 'related, either directly or indirectly, to some aspect of CFC manufacture or usage' (DoE 1976: 6). In a 1980 communication (COM(80)339 16.6.80), meanwhile, the Commission stated that any further reduction in CFC use in aerosol sprays (beyond the 30 per cent figure) would 'be likely to cause socio-economic problems because of existing overcapacity in the industry' (Haigh 1989: 267).

References to the United Kingdom in the above paragraphs suggest that this country had a particular interest in the development of the debate. And so it did. At the end of the 1970s, the UK was reported to be the largest producer of aerosol units in the Community (Haigh 1989: 267), and British entities worked to defend these interests. In 1978, for example, a report published by the British Aerosol Manufacturers Association noted that there was much scientific confusion, and therefore 'there is no hazard in waiting for more definite scientific conclusions, nor any reason to restrict manufacturers' choice of propellants for aerosol cans' (NS 1978: 830). The British government often echoed such views.[8]

Thus, there is some strong evidence to suggest that there was a correlation between concerns for competitiveness and external policy on this particular environmental issue during the late 1970s. Indeed, Richard Benedick – who served as the lead negotiator for the United States during much of the negotiations – is thus convinced. He argues that during the 1970s, the EC Commission:

> was sympathetic to industry arguments that strong controls on aerosols would impose hardships because of the substantial existing overcapacity and the allegedly large capital investment required to

convert to the hydrocarbon propellants used by companies in the United States. . . . In addition, all the EC 'regulations' were actually implemented by voluntary agreements with the manufacturers. In sum, these were painless moves, fully supported by European industry, that gave an appearance of control while in reality permitting continued expansion.

(Benedick 1991: 25)

With a change of administration in the United States, coupled with less urgent scientific reports, the ozone layer debate fell down the international agenda during the first half of the 1980s (Rowlands 1995a). Notwithstanding the opening for signature of a framework convention in March 1985 (see Table 2.1), less pressure existed for further regulation. This attitude, however, changed dramatically with the announcement of the discovery of an 'ozone crater' over Antarctica in May 1985, and more urgent scientific findings.

Protocol negotiations, 1986–1987

With this dramatic, and unexpected, finding, the international negotiations were given a greater sense of urgency. Informal discussions about an international protocol took place throughout 1986; the following year, formal negotiations began. (Note, in Table 2.4, that by 1986, EC production of CFCs had clearly surpassed that of the United States, making the Community the largest single producer in the world.)

The US led the calls for strong controls, continuing to argue the case for an immediate prohibition on CFCs as aerosol propellants. In light of this particular emphasis, the European reaction continued to be cautious. For his part, Benedick argues that:

Table 2.4 Global CFC production by country/region, 1986

Country/Region	Percentage share
European Communities	42
United States	28
Japan	12
USSR and Eastern Europe	11
Developing Countries	4
Other OECD	3

Sources: Compiled from Gamlen (1986), Thornton (1990) and UNEP (1993)
Note: All calculations by weight

Some European industrialists had suspected all along that the United States was using the ozone scare to cloak commercial motivations. They now believed that American companies had endorsed CFC controls in order to enter the profitable EC export markets with substitute products that they had secretly developed.

(Benedick 1991: 123).[9]

Although this position might be a bit extreme,[10] concerns about the impact of the international negotiations upon the economy were invariably affecting the EC negotiators' perspective. However, they recognised that if they were not at least seen to be receptive to the Americans' efforts to forge an international agreement, then the United States could well follow up their fresh threat to take unilateral action and impose trade sanctions (Haas 1993: 165). For this and other reasons, some action was considered.

For the purposes of this chapter, it is appropriate to focus not only upon the extent of reductions in CFCs being advocated by different groups in the negotiations during the first half of 1987, but also upon the kinds of reductions. The latter highlights some differences of opinion. On the one hand, the United States (and its allies, often collectively referred to as the 'Toronto Group')[11] continued to push for a control upon particular end uses (specifically, the can ban). The EC, on the other hand, wanted attention directed to all CFCs, arguing that a molecule of CFC had the same impact upon the ozone layer, irrespective of its source. Their representatives also argued that production – rather than consumption – should be the figures that were controlled, for this would be easier to monitor. These preferences can be explained at least in part by the impact that specific end use controls upon consumption would have upon EC competitiveness.

Because of the already existing ban on CFCs as aerosol propellants in the United States (and some other countries), alternatives (both substitute chemicals and new technologies) had been developed by American industry. Consequently, if a can ban were at that time to be implemented world-wide, then American companies would have potentially lucrative export markets opened up by a global ban. Moreover, even if there were regulations placed upon all CFCs (that is, not distinguishing between end-uses), the US would still have an interest in advancing a 'consumption' basis. If this were the case, then CFC exports would effectively escape regulation (for consumption would presumably be calculated by production minus exports plus imports). Faced with no restraints upon future international activity, the large US producers of CFCs could be tempted to compete against European producers around the world.

Alternatively, however, if the regulations limited production, then the dominance of the EC in global export markets would be preserved (for any other country could only increase export sales at the expense of domestic consumption). And, at a much more basic level, restrictions that did not focus upon end-uses would also favour the EC: they could make the initial

41

reductions in areas of their choice. Alternatively, a global can ban would not only serve to cut off valuable export markets, but could also force EC countries to import alternative chemicals and replacement technologies. Obviously, this would be a double blow to the member states' trade balances.

Not surprisingly then, the EC, even when it accepted the notion that cuts were possible, was still only willing to consider such moves for CFC production as a whole – ostensibly for two major reasons. First, because it would be easy to administer (because of the relatively small number of CFC producers), and second, because any ban on particular end-uses would only be a temporary remedy (for increases in other uses would inevitably continue). Nevertheless, it is clear that concerns about competitiveness inevitably were also a part of the calculation made by EC representatives.[12]

In the end – that is, at the conference in Montreal in September 1987 where a protocol was eventually agreed – a compromise was reached: both production and consumption were controlled, with the former permitted to be slightly higher than the latter. (This was designed to create excess supply in northern countries, which would thus allow a limited expansion in exports in order to meet the 'basic domestic needs' of developing countries.) Given this – along with the lack of differentiation among end-uses in the Protocol – it is clear that the 1987 Treaty reflects EC economic interests, at least to some extent.

The view that the Community's actions during the Protocol negotiations can be explained, at least to a significant degree, by concerns for competitiveness and industrial interests more generally, is one that is held by many commentators. Karen Litfin, for one, argues that during 1986, the 'EC's position was strongly influenced by industry; in fact, industry representatives sat on the delegations of some EC countries' (Litfin 1994: 107). Benedick also maintains that the EC view 'followed the industry line and reflected the views of France, Italy, and the United Kingdom' (Benedick 1991: 68). Oye and Maxwell argue that both 'the US and the EC wished to prevent the other's industry from gaining a competitive advantage through the content of an international agreement that limited the usage of CFCs' (Oye and Maxwell 1995: 199). And perhaps the most telling indictment is a quotation from the individual who was shepherding the international negotiations at this time, UNEP Executive-Director Mostafa Tolba, who reportedly said that the

> difficulties in negotiating the Montreal Protocol had nothing to do with whether the environment was damaged or not. It was all about who was going to gain an edge over whom; whether Du Pont would have an advantage over the European companies or not.
>
> (MacKenzie 1988: 25)

Notwithstanding these observations, however, it is clear that the entire process cannot be wholly explained by concerns for competitiveness and industrial interests more generally. Pressure for regulatory reform derives from advances in scientific knowledge, competing goals of global environmental protection, and the dynamics inherent to the bargaining process itself. Indeed, simply note that the Montreal Protocol was actually agreed, and that it imposed significant obligations upon its parties: before conclusive scientific evidence was available, countries had agreed to halve their production and consumption of a range of 'vital' industrial chemicals. Moreover, Edward Parson argues that concerns for competitiveness found less voice as the negotiations continued: 'Several observers contend that it was sometime between September 1986 and April 1987 that European industry lost control of national delegations' (Parson 1993: 42).

Finally, telling the story as we have done here suggests a certain uniformity of purpose and interests on the part of the EC and its member states during the negotiations towards a Protocol. This was certainly not the case, for it is very important to highlight the differences *within* the Community – that is, the way in which different perceptions of economic impacts did or did not play a role in the policies of various member states. In such a discussion on the Protocol negotiations, attention is usually initially directed to the United Kingdom and France.

Table 2.5 reveals that the United Kingdom was Europe's second-largest producer of CFCs in the mid-1980s.[13] Within this, meanwhile, Imperial Chemical Industries (ICI) was the major British producer.[14] The chemical industry in general, and ICI in particular, has traditionally had a special position in the eyes of successive British governments.[15] As a consequence, many argue, ICI exercised a considerable influence upon UK decision-making, pushing for a sceptical 'go slow' approach. During the late 1970s and early 1980s, for example, while some were pushing for a

Table 2.5 CFC production in EC member states, 1986

Country	Tonnes of CFC produced	Country	Tonnes of CFC produced
Belgium	0	Italy	60,000*
Denmark	0	Luxembourg	0
France	73,157	Netherlands	45,859
West Germany	125,579	Portugal	0
Greece	14,040	Spain	34,896
Ireland	0	United Kingdom	106,129

Sources: UNEP (1993), Greenpeace (1995)
Note: * = estimate

wider negotiation mandate for the Commission, the British (often with the French) ensured that the Council, which had to decide unanimously, was not granted it (Jachtenfuchs 1990: 265). For his part, Benedick argues that ICI not only influenced British policy, but was also 'the driving force in the European Council of Chemical Manufacturers' Federations, which was an active lobbyist in Brussels and a conspicuous presence during the negotiations' (Benedick 1991: 39). Indeed, the movement that was eventually seen in the EC's negotiating position (particularly during the summer of 1987) stemmed significantly from a relaxation of the British attitude. This, in turn, many argue, came about as a result of a change in attitude on the part of ICI (Jachtenfuchs 1990: 268, Maxwell and Weiner 1993).[16] Thus, the concern for economic competitiveness and industry was prevalent in the British position.

A similar story can be written about the influence of Atochem upon France, the Community's third-largest CFC producer in 1986. The interests of the French government in Atochem extended beyond simple concern about the health of the nation's economy as a whole: it is part of the state chemicals giant Elf Aquitaine. As a result, the French government – and occasionally individual Commissioners[17] – worked to protect their interests. It was only after the British *volte-face*, when the French found themselves virtually isolated in their belief that controls could be slower, that a relaxation of policy was considered.[18]

Turning to the largest producer of CFCs in the Community, one might similarly expect industry to influence the position of government and to advocate a 'go slow' approach. This, however, was not the case. West Germany was more pro-active in the negotiations than the other major Community producers. Indeed, Benedick credits West Germany with being 'at the forefront' in the drive for tighter controls on CFCs (Benedick 1991: 113).[19] In 1986, for example, the country registered a formal protest when a Council Decision was, in its eyes, too conservative (ibid.: 69). Additionally, during the negotiations for the Montreal Protocol, the German representatives pushed for an aerosol ban and a 50 per cent overall reduction (ibid.: 84). They also co-sponsored reports (for example the 1986 WMO/NASA Assessment) and held a number of major scientific meetings on the issue (Litfin 1994: 110).

The West German calls for regulatory reform on ozone layer depletion can largely be explained by domestic public pressures. Indeed any belief, on the part of that country's decision-makers, that the demand for controls on CFCs could be resisted, were discarded after the 1986 elections – a time at which the Green Party made an impressive showing. Because of this tide, officials in both the West German government and the country's chemical industry had a desire for Europe to take the lead on the issue, for a variety of reasons. The businessperson, for one, wanted 'a common set of conditions for marketing [his/her] products in order to avoid costly duplication of tests' (Grant, Paterson and Whitson 1988: 289). Second, regulation at the European level would more likely be weaker than it would otherwise be if made only in West Germany. As Grant and colleagues argue, West German industrialists recognised that, at the European level,

> the 'Greens' are weaker, and other countervailing forces can be mobilised. At the very least, it can be ensured that the 'misery is shared' so that the German chemical industry is not disadvantaged by more stringent environmental regulation than applies elsewhere in Europe.
>
> (Grant *et al.* 1989: 78)

Third, West German support for EC solutions is also fuelled by 'a high degree of compatibility between their political/administrative assumptions and community procedures' (Grant *et al.* 1988: 290). Thus, West German willingness to adopt a strong European position might still be explained, at least in part, by concerns about economic competitiveness and the health of industry more generally.

Nevertheless, the caution that was called for when examining the Community as a whole should also be exercised here. There were costs upon all EC countries. Even in those that were not producers of CFCs, there were entities – for example, aerosol propellant producers, dry-cleaners, refrigeration, etc. – that were reliant upon the use of CFCs in the goods that they produced and sold. Conceivably, therefore, all member states could be adversely affected by any restriction upon the production and consumption of CFCs. In spite of this, a number of countries were extremely forward looking. Denmark, for one, pushed for a strong consumption formula during the protocol negotiations (Benedick 1991: 80). The Netherlands and Belgium were two other countries often identified as part of this more 'pro-active' grouping (ibid.). The negotiating positions of these states highlights again the occasional dominance of competing sources of pressure for regulatory reform.

Accelerated phaseout, 1987–1995

Following the Protocol negotiations, implementation in the Community was the first task. Under the West German presidency during the first half of 1988, that government's representatives tried to push the Community to go beyond the terms of the Protocol (and thus implement even tighter controls upon ozone-depleting chemicals, as the West Germans had done unilaterally). Though these efforts helped ensure the Council agreed to ratify both the Vienna Convention and the Montreal Protocol in June 1988, Jachtenfuchs argues that the wishes of DGIII (internal market) were able to reign supreme (over not only the preferences of the West Germans, but also DGXI (environment)), and the Protocol was implemented virtually 'word-for-word' in EC regulation (See Table 2.2).

Nevertheless, by the end of 1988, a further change of attitude took place among the traditionally sceptical in the Community. With the 'greening' of Mrs Thatcher in the autumn of 1988, the door was opened for a more pro-active British position. The final thrust for this came not only from the impact of British scientific reports,[20] but also the agreement from ICI for a total phaseout.

Though the final shift in the ICI position is publicly attributed to an industry review of the Ozone Trends Panel's report from March 1988 (Parson 1993: 46), Maxwell and Weiner place emphasis upon commercial considerations. They note that in 1988 it was expected that CFCs would largely be replaced by other chemicals. As a result, opportunity for first-mover advantages arose: '[P]roducts and market share would accrue to the companies that developed process technology for making substitutes in the most cost effective and rapid manner possible' (Oye and Maxwell 1995: 200). The fact that the substitute chemicals might be five times as costly as CFCs lent further impetus (Maxwell and Weiner 1993: 33–34). Finally, Litfin high-lights the attraction for ICI of potential international markets – that is, if global controls were implemented, then global markets for substitutes could well be forthcoming. Ironically, 'ICI paid the travel expenses for some dele-gates from developing countries to attend [the March 1989 conference in London on "Saving the Ozone Layer"]' (Litfin 1994: 211–212, fn 8). The French change of opinion, meanwhile, was further prompted by increased isolation, and desires on the part of President François Mitterand to do nothing to blemish his environmental credentials.[21] The interests of Atochem in the potential substitute chemicals (see below) no doubt also played a role.

So by the end of the decade it was clear that the days of CFC use in the industrialised world were numbered.[22] While this is not meant to suggest that the challenge posed by CFCs had necessarily been met – developing country use, illegal trade and implementation more generally are still three outstanding tests – it is meant to imply that the most contentious debates about CFC production and use in the European Union had, by the end of the 1980s, been resolved.

Controlling HCFCs

With the elimination of CFCs gaining widespread support among the OECD countries during the late 1980s, greater political attention could be paid to the other potential ozone-depleters. Top of this list were hydrochlo-rofluorocarbons (HCFCs). HCFCs were identified as potential substitutes for up to 30 per cent of the controlled CFCs.[23] Although more ozone-benign than CFCs, they were by no means 'wonder chemicals', for they still destroyed some stratospheric ozone. In light of this fact, pressure mounted during the first half of 1990 to impose restrictions upon their use at the Second Meeting of the Parties in London in June of that year.

A number of industry officials resisted this pressure. They maintained that society's primary goal must be to eliminate CFCs as quickly as possible. Although there were problems associated with the use of HCFCs, they were perceived to be the most attractive alternative, at least during this 'transition phase'. Thus, the argument continued, it was better to use the lesser of two evils while more appropriate alternatives were being developed and tested – in this instance, therefore, the ends justified the means. This view generally prevailed during the June meeting, and no restrictions were placed upon the chemicals. Instead, signatory nations called for 'producers to use [HCFCs] responsibly, and work towards eliminating them by 2040' (Schoon 1990: 1).

Identifying this as the industry view, however, hides some interesting variations among companies' respective interests in HCFCs (see Table 2.6). Karen Litfin, for one, reports that the German CFC industry 'was silent on the issue of HCFCs, having chosen not to invest in chemicals that would only be transitional and to rely instead on ozone-safe HFCs [hydrofluorocarbons]' (Litfin 1994: 150). In the United Kingdom, ICI had taken a similar strategic decision – deciding not to invest heavily in HCFCs, but instead also to go on to HFCs (Maxwell and Weiner 1993: 34–35). As a consequence, both entities had interests in regulatory reform which placed controls on HCFCs – or at least for clear signals to be sent that HCFC regulation of some kind was inevitable – for this would encourage users to leapfrog the use of HCFCs and proceed straight to HFCs.

By contrast, the French company Atochem had taken a strategic decision to invest heavily in HCFCs. Between 1986 and 1995, the French were responsible for 40 per cent of EC production, more than twice as much as the second-placed British. Indeed, one report suggested that Atochem was, in 1992, the largest single producer of HCFCs in the world (ENDS 1992a: 14).[24] One might therefore expect the French to be more resistant to controls upon HCFCs. This indeed has been the case. As a consequence, the Commission was prevented from taking a pro-active line on HCFCs during

Table 2.6 Estimated HCFC production, 1986–1995

Country	Tonnes of HCFC production	Country	Tonnes of HCFC production
Belgium	0	Italy	68,400
Denmark	0	Luxembourg	0
France	335,000	Netherlands	95,000
Germany	95,000	Portugal	0
Greece	20,000	Spain	75,000
Ireland	0	United Kingdom	155,000

Source: Greenpeace (1995)

the negotiations at the Fourth Meeting of the Parties in Copenhagen. While most of the member states wanted the CFC element of the HCFC cap to be set at somewhere between 2 and 2.5 per cent, the French still wanted it to be pegged at 4.0 per cent (ENDS 1992a: 14).[25] French resistance was buttressed by both the United States and the developing world.

The position of the US delegation was informed by the fact that HCFCs are used in the large air conditioners that cool office buildings. The country that makes the most use of these large chillers is the United States. Because these air conditioners have economic lifetimes of up to forty years, business people wanted to ensure that they would be able to keep these chillers operational throughout this period. The 0.5 per cent usage allowed between 2020 and 2030, which the US delegation demanded and received, guaranteed these machines' continued utility.

Developing country representatives, meanwhile, were concerned that a quick contraction of the OECD market for HCFCs would have knock-on effects for their own development prospects – that is, this possible substitute would not be available as an option. Wanting no possibilities to be prematurely, and unfairly, foreclosed, they also pushed for laxer controls. Given this, the majority of the EU countries could do little to prevent the cap from being set higher than they would have liked – namely, 3.1 per cent (Rowlands 1993).

When EC Environment Ministers met the following month (December 1992) in order to consider how the changes agreed at Copenhagen might be implemented (and whether or not the Community should go beyond the global goals), this debate continued. While some member states – particularly Germany and Denmark – pushed for an accelerated timetable (phaseout by 2000–2005) and a lower cap, the French continued to lead the resistance to this. During the beginning of 1993, this debate proceeded behind the scenes, with the French continuing to press for nothing more than the terms of the Copenhagen agreement (EB 1993: 4). Indeed, though DGXI had a proposal ready for the Environment Council which went beyond the terms of the Copenhagen agreement, this effort was apparently blocked in March by Commission President Jacques Delors, who was reportedly following objections raised by the French company Atochem (ENDS 1993a: 37).[26]

Three months later (in June), a proposal was finally tabled at a meeting of the Council. In this, the magic figure for the cap was set at 2.5 per cent (as compared with the 3.1 per cent agreed at Copenhagen), and the complete phaseout date proposed as 2014 (as compared with 2030) (ENDS 1993a: 37). The respective groups of supporters and opponents that formed in response did produce one surprise. As expected, Denmark and Germany continued to push for more – each citing its own domestic plans for going well beyond the terms of either the Copenhagen amendment or even the draft proposal. Equally as expected, the French delegation said that it was opposed to any new controls upon HCFCs. What was surprising, however, was that the Italians 'not only agree[d] to the tighter HCFC controls proposed in the regulation but also hint[ed] at possibly going further' (ENDS 1993a: 38).

Explaining this movement, participants suggest that although some of the southern European countries were instinctively opposed to further controls on HCFCs, they feared that their opposition might be used against them in the methyl bromide debate. More specifically, given the demand on the part of the United States for strong controls on methyl bromide (for reasons outlined in the next section of this chapter), the southern Europeans believed that they needed a stronger position on HCFCs (the issue on which the US was the dragger) in order to ensure that their methyl bromide position remained tenable. In other words, they were willing to 'budge' on HCFCs to justify a 'stick' on methyl bromide.

After another couple of months, EC Environment Ministers finally agreed something close to this proposed regulation: a 2.6 per cent cap and a 2015 phaseout date (AE 1993a). Reports suggest that this was only decided after qualified majority voting, with the best guess being that the French and the Spanish (who had an Atochem-owned HCFC plant on their territory) together constituted the defeated opposition (ENDS 1993b: 35).

The European Parliament, however, was not satisfied with this, and proposed its own set of amendments. They wanted the cap to be lowered to 2 per cent and HCFC consumption to be eliminated by the year 2002. The parliamentarians also proposed that controls be placed upon production as well as consumption (with attendant implications for HCFC exports) (ENDS 1994a: 36). These amendments, however, were rejected by the Ministers when they reconvened in June 1994 (ENDS 1994b: 36). The regulation received a second reading from the Parliament in the late autumn of 1994 and entered into force on 23 December 1994.

With the tenth anniversary of the Vienna Convention approaching, attention turned to the possibility of regulatory reform which tightened *global* controls on HCFCs yet again. The EC Environment Commissioner Ritt Bjerregaard called, in September 1995, for greater action in a range of areas, HCFCs included. A number of countries – Austria, Belgium, Denmark, Finland, Germany, the Netherlands and Sweden, of course backed by the Commission itself – picked this up and pushed for a strong negotiating position at a Council Meeting held on 6 October 1995. Others – France, Greece, Ireland, Portugal and Spain – felt less of a sense of urgency. Nevertheless, the former were successful, securing a relatively forward looking negotiating position for the Commission to take to the Vienna Conference of the Parties two months later. This position called for the cap to be brought down to 2 per cent, leaving the phaseout date at the already-agreed date of 2015 (EE 1995: I1). Again, the fact that the Union was making little progress on the methyl bromide issue helps to explain this relatively pro-active stance.

In Vienna in December, the international community reached a compromise on HCFCs. While the reduction and elimination timetable remained the same (primarily at the continued insistence of the United States), a lower baseline figure was accepted – led by the EU, the new controls used as a base 2.8 per cent of 1989 CFC consumption plus 1989 HCFC consumption. Critics continued to highlight the fact that controls were placed only upon consumption, not production as well. As a consequence, the potential for export production in the EU remained high. Indeed, given that the controls on HCFC consumption in developing countries were rather far in the future – a stabilisation date of 2016 (based on 2015 levels), with a final phaseout by 2040, was agreed in Vienna – the scope for massive increases in HCFC production remained.[27] This no doubt pleased those with large HCFC production capacity.

Indeed, the development of the European position can, in part, be explained by concerns about industrial interests. With the French so heavily committed to HCFCs as alternatives to CFCs, it is understandable that they would want the Union to go slower on proposals for controls. This is given greater tangibility when one recognises that a 0.1 per cent difference in the cap calculated (that is, for example, the difference between a cap at 2.5 per cent and one at 2.6 per cent) would mean a difference in allowable EC consumption of approximately 3,500 tonnes a year,[28] which could translate into sales revenues of US$10 million a year.[29] Though this annual figure is by no means astronomical, it could easily grow into significance over a period of years. Accordingly, the interests of the French in securing as long a payback period as possible are not surprising.

Ongoing concerns

Controlling methyl bromide

Just as HCFCs began to attract considerable attention at the 1990 Meeting of the Parties, methyl bromide garnered many of the headlines at the 1992 Meeting. In the EU, this ozone-depleting substance is mainly used as a soil disinfectant in agriculture, as well as a disinfectant for agricultural products. The only methyl bromide producer in the Union is Atochem (ENDS 1993c: 37), with Greenpeace estimating that it made 50,000 tonnes of the chemical between 1986 and 1995 (just over 7 per cent of the global total). Use of the compound in Europe, meanwhile, has been estimated at 19,000 tonnes per year (about one-quarter of the global total) (van Haasteren 1994: 16).

In preparation for the 1992 Copenhagen Meeting of the Parties, European Environment Ministers met on 20 October of that year. At that meeting, the Dutch representatives pushed for tight European regulation and a strong negotiating position at the upcoming Conference of the Parties. Their experience with the chemical had been far from positive: because it had polluted soil and groundwater in horticultural areas, the Dutch government had already agreed to ban it at home. As a first step towards a similar Europe-wide policy, they urged a 70 per cent reduction by 1995 (ENDS 1992b: 37).

Others, though less adamant, still called for regulation. The United Kingdom, for example, supported a 25 per cent cut in methyl bromide production by the year 2000. However, a number of southern states (namely, Greece, Spain and Italy – three major users of the chemical – and France – the major producer) fought these moves (ibid.). The interests of these countries, particularly the consumers, lay in the fact that methyl bromide was such an effective fumigant (it is easy to apply and it causes yields to increase dramatically) and that the agricultural crops most dependent upon it (strawberries and tomatoes) were important to their economies. They were particularly concerned about how competitors across the Mediterranean might be able to gain an advantage, were the use of methyl bromide to be restricted in their home countries. After debate and discussion, the EC Ministers agreed that the 'Community would ask the Copenhagen meeting to include methyl bromide for the first time on the list of controlled substances, seeking to stabilise output at 1991 levels by 1995' (Reuters 1992).

Amid much debate over methyl bromide in Copenhagen, the more sceptical among the European countries highlighted uncertainties in the special assessment reports to support their position. They were supported in this position by many developing countries, who feared not only increasing prices due to restricted supplies, but also the prospect of trade restrictions upon their agricultural exports (which are in effect 'made with' the chemical) (Rowlands 1993: 27–28).

On the other hand, the United States was among the most pro-active, originally calling for the elimination of non-essential uses of methyl bromide by the year 2000 (UNEP 1992: 18). United States interest in the issue was focused by the fact that, because it was anticipated that the ozone-depleting potential (ODP) of methyl bromide would be set at 0.7, the United States would automatically be required to take action, regardless of the outcome of the Copenhagen Meeting. Under the terms of Title VI of the US Clear Air Act (1990 Amendments), the production and importation of any substance with an ODP above 0.2 must be phased out within seven years. Given the importance of agriculture to states like California and Florida – and the potential of Mexican farmers to gain a competitive advantage (for they could continue to use this effective fumigant) – the US delegation pushed hard for global controls. They were only partially successful, however: it was finally accepted that industrialised countries would freeze their consumption at 1991 levels by 1995 (ENDS 1992c: 34).

After Copenhagen, the Commission emerged with a proposal on the way in which the Conference's commitments could be implemented by the Community. Going beyond simply what was agreed in Copenhagen, reports circulated that the Commission might propose a 20–30 per cent cut by the middle of the decade, with further reductions towards the decade's end

(ENDS 1993d: 38). Were there ever such a proposal, it fell victim to the same forces that brought down the HCFC proposals (see above), and thus what emerged in June 1993 was tamer.[30] Nevertheless, it still went further than the international agreement, for the Commission proposed a 25 per cent reduction of 1991 methyl bromide production levels by 1996 (though no further cuts were mentioned).

At a meeting of the Environment Council in June 1993, France and Greece immediately registered their opposition to the Commission's recommendations; Italy reserved its previous position in opposition, and some observers suggested that Spain was wavering (ENDS 1993a: 38). However, at a Council meeting on 5 October of the same year, some movement was evident. More specifically, France and Spain suggested that a 25 per cent reduction was allowable, with it taking effect in 2000 (ENDS 1993b: 35).

When the Regulation was finally agreed by ministers at the end of 1993, consumption levels had been frozen at 1991 levels from 1995 (mirroring the international position), while a 25 per cent reduction by 1998 was to follow (thus going beyond the international position). This was a compromise proposal that had been put forward by the Belgian presidency, meant to fall between those who wanted more (including Denmark, the Netherlands and Germany, many of whom were taking unilateral action on the chemical) and those who wanted somewhat less (particularly France, Greece, Spain and Portugal) (ENDS 1993c: 37).[31]

During preparations for the 1995 Vienna Conference, however, methyl bromide again attracted much debate. When discussions in the Council began, it became clear that the dividing lines had shifted slightly, though perhaps significantly. While Greece and Portugal continued to oppose any further controls, two traditional opponents of regulation – Spain and France – indicated their willingness to support a 50 per cent cut by the year 2005.[32] By doing so, they joined Italy in a common position (ENDS 1995a: 42, 1995b: 37). On the other hand, the Nordic countries, the Netherlands and Austria pushed for a complete phaseout by 2001 (EE 1995: I1). The United Kingdom and Belgium took an intermediate position: supporting a 50 per cent reduction by 2000 (ENDS 1995a: 42, 1995b: 37).

In the end, the position of the Chair – that is, Spain – prevailed, and with this stance, the EU entered the Vienna negotiations. At the Conference of the Parties itself, methyl bromide proved to be the key issue upon which the Union was 'behind' most other major industrialised countries. (See the discussion in the previous section about how this issue is linked to the HCFC debate.) Continuing to be spurred by the economic interests generated by its own domestic legislation, the United States pushed for an advanced phaseout date (though with significant exemptions).[33] The EU, however, tried to resist this movement: though Germany and the UK reportedly supported a 2001 phaseout, southern states (particularly Spain) desired less (EPL 1996: 67). In the end, the Parties agreed to a complete phaseout of methyl bromide by 2010. Reflecting its concerns for a level playing-field,

> Spain, acting as EU President, insisted that the process should include trade measures against non-parties to the Protocol. Its main concern is that it will suffer from unfair competition from Morocco which is not a party and will be able to use methyl bromide freely.
> (ENDS 1995c: 36).[34]

Again, we see concerns about competitiveness entering discussions about the EU's external environmental regulatory reform policy on ozone-depleting substances.

Other issues

The focus upon CFCs, HCFCs and methyl bromide should not distract from the fact that there are other ozone-depleting substances. Controls on halons were present in the original Montreal Protocol, while regulations on carbon tetra-chloride and methyl chloroform were introduced in 1990; the elimination of hydrobromofluorocarbons, HBFCs, moreover, was agreed in 1992. It is certainly the case that concerns about industrial interests, and competitiveness more generally, have affected the discussions about controlling the production and consumption of these chemicals. The United Kingdom, for example, pushed for a slower phaseout of methyl chloroform during a debate among European ministers in early 1992. It was reported that the UK position was 'on the grounds that it is used by many small businesses which face difficulties in finding substitutes' (ENDS 1992d: 68). The fact that ICI was producing it as a substitute for CFC-113 no doubt also entered the calculation.

There are other issues that have been, and will probably continue to be, important to this debate. The definition of 'essential uses' is one such instance. Companies and governments, for example, may use this as a means to protect potentially affected interests and entities. The concern about the production and consumption of ozone-depleting chemicals in the developing world is another. We have already seen this interest in the debate about HCFCs (with European producers wanting to ensure that markets for these chemicals will emerge in the South). Any movement to bring forward controls on HCFCs would inevitably generate even further debate.

Concluding remarks

The purpose of this chapter has been to explore regulatory reform in the sector of ozone-depleting substances, highlighting how the conflicting goals of environmental protection and economic competitiveness influenced policy change. Concerns about economic competitiveness, and the interests of industry more generally, have affected the development of the European

Union's policy on ozone layer depletion. From the views expressed by the British over CFC regulation, through the interest shown by the French in HCFC controls, to the position of the southern EU countries in the methyl bromide debate, it is clear that concerns about the way in which an EU policy may impact member states' interests, economically-defined, has played a role in EU decision-making: states resisted reforms which could undermine their domestic economic positions, and advocated reforms which either expanded their export markets or preserved their competitiveness by spreading internationally the costs originally borne through unilateral action. Moreover, these concerns have expressed themselves both inside the Union – that is, in the Council of Ministers – and externally – that is, when the EU has participated in global negotiations. Consequently, those advancing the thesis that industrial interests and worries about economic competitiveness exercise an important influence on the formation of environmental policy in the EU have been given some support by this case-study.

At the same time, however, the evidence has also revealed that there were a number of instances throughout the history of the issue when regulatory reform could not be explained by corresponding shifts in economic calculations – for example, the relative greening of the United Kingdom in 1988/89, as well as the incremental change of heart shown by the French in the early years of the present decade. Caution, therefore, should be exercised when striving to explain the development of EU environmental policy. Other factors – for example, non-economic concerns on the part of member states, the functioning of domestic institutions, the effects of advancing scientific knowledge or simple conference dynamics – must have played a part.[35] Indeed, this chapter has illuminated occasions when the Commission, acting as a semi-autonomous agent of environmental protection, successfully pushed regulatory reform faster or in a different manner than the states desired.

Therefore, a cautious endorsement of the hypothesis – that is, that concerns about economic competitiveness, and the interests of industry more generally, have influenced the development of the European Union's policy on ozone layer depletion – is offered in this chapter. It certainly helps to make sense of many of the intra-European and global debates and negotiations on the issue of ozone layer depletion, but it is not a panacea.

Notes

1 The author thanks Andrew Jordan (CSERGE) and the participants at the workshop in Florence for their helpful comments on an earlier version of this chapter.
2 For further information about the relationship between 'environment' and 'economy', see WCED (1987) – the publication that did more than any other to popularise the notion of 'sustainable development' – and MacNeill *et al.* (1991).
3 Environmental outcomes at the EC level have usually taken the form of 'Directives', which, though binding on all member state with respect to the result to be achieved, nevertheless leave responsibility with regard to the form

and means to be used to achieve the stated objectives up to the member states. Alternatively, the legal form for implementation of the Montreal Protocol and its amendments has been the 'Regulation'. Regulations have direct application in the member states and come into force immediately. Jachtenfuchs argues that this was done 'in order to avoid trade distortions resulting from non-simultaneous application of the proposed legislation, and also to underline the urgency of the matter' (Jachtenfuchs 1990: 269).

4 The issue of external competence and the Commission's ability to negotiate on behalf of the member states also plays a crucial role in the cases of pesticides (Chapter 3), marine pollution (Chapter 4) and carbon emissions (Chapter 5).

5 Some of the discussion in this section is taken from Rowlands (1995a, Chapter 5).

6 A full description of the US position may be found in Benedick (1991).

7 In 1973, approximately three-quarters of all CFCs used in the EC went into aerosol sprays, while in the US, the figure was just under one-half. This also meant that, even in spite of the greater consumption of CFCs in the United States, there were still more aerosol units, in absolute terms, filled with CFCs in the Community as compared with the United States. (Data from OECD 1976: 14, Engelmann 1982: 53 and Layman 1986: 42.)

8 In April 1976, UK Environment Minister Denis Howell remarked: 'To put this matter [potential ozone layer depletion] in perspective, this [any projected increase in ultraviolet radiation reaching ground level] would be equivalent to the increase in exposure incurred by a person moving from Northern England to the South Coast of England' (HC 909 567 W, quoted in Purvis 1994).

9 Benedick goes on to note that: 'this suspicion was unfounded: to the dismay of environmentalists, du Pont had admitted in 1986 that it had ceased research on chlorofluorocarbon alternatives in 1981' (p. 123).

10 For the argument that Benedick misunderstood and therefore misrepresented key elements of the European position, see McConnell (1991).

11 This included the United States, Canada and the Nordic countries.

12 Additionally, the EC insisted upon being treated as a single unit for the purposes of consumption. Its representatives argued that, because this figure included imports and exports, restrictions upon the consumption levels of individual member states would pose a barrier to trade within the Community and thus contrary to the Treaty of Rome and efforts to create a Single Market (Benedick 1991: 95–97).

13 Any suggestion as to production levels of chemicals must be taken cautiously. Owing to concerns about inter-industry secrecy, companies are loath to publish explicit figures about production levels (or the value of the same) of chemicals. Consequently, such figures are often estimates or even 'best guesses'.

14 'By the late 1980s, ICI accounted for more than 80 percent of total British CFC output' (Maxwell and Weiner 1993: 33).

15 See, for example, Grant *et al.* 1989: 78. Moreover, Maxwell and Weiner argue that 'the British departments were unapologetic about the closeness of government–industry collaboration' (Maxwell and Weiner 1993: 30).

16 ICI were also concerned that '[C]onsumer-driven environmental concerns, which were widespread in 1987, were not limited to CFC products, and could potentially diminish the sales of the entire ICI Group' (Maxwell and Weiner 1993: 33).

17 Benedick – developing his general thesis that the European view was informed by a need to protect industry – reports that a paper distributed by the French government, with 'EC views' contained therein, was published on Atochem letterhead (Benedick 1991: 78).

18 The Spanish and Italians have also been identified as 'draggers' by some.

19 He also notes that West Germany 'was instrumental in turning around the European Community's initially negative position' (p. 6).

20 The UK's Stratospheric Ozone Review Group released its report in the summer of 1988. In this, they agreed that the controls in the Montreal Protocol would not be sufficient to prevent further significant destruction of the ozone layer (UKSORG 1988).

21 The French change of heart (which was fully apparent by March 1989) was explained thus: 'France, maintaining that unilateral reductions would not save the ozone layer and would only benefit non-European companies, had found itself isolated in its opposition to stringent reductions. But France relented, apparently favouring its political interests over its economic interests. President Mitterand, taking a leadership role in efforts to prevent global climate change, was cosponsoring an international conference on the global environment in The Hague later that month. Had his government blocked efforts in the EC to save the ozone layer, Mitterand would have been in an embarrassing position at that conference' (Litfin 1994: 128–129).

22 Many CFC users, for their part, were not resisting further controls, for they eventually found that alternative methods were often cheaper than the CFC-based procedures they replaced.

23 This was the estimate at the time. It has since been revised, for hydrocarbons have taken much more of the market than originally anticipated.

24 A Greenpeace report suggests the same for the period 1986–1995 (when taking ozone-depleting potentials of different HCFCs into account) (Greenpeace 1995).

25 At Copenhagen, the French also broke ranks by pushing for the intermediate target (that is, a 35 per cent reduction) to take place in 2010 rather than 2003, as previously agreed by the EC (Follett 1992).

26 A spokeswoman for Atochem called such charges of interference 'idiotic stories'. However, a Commission source, speaking on the condition of anonymity, claimed: 'You can quite rightly blame France. . . . They're going to attack [the Commission's proposals] so that Atochem will be able to get the greatest possible (HCFC) consumption in the EC' (Love 1993). It was reported later that the industry and agriculture directorate generals, DG III and DG VI respectively, also hindered the progress of the Commission's proposal (Reuters 1993).

27 Indeed, these controls may serve to encourage short-term consumption of these chemicals in the developing world (in order to increase a country's base-line figure). See, generally, Krueger and Rowlands (1996).

28 This calculation is based on the fact that 1989 consumption of CFCs in the EU-15 was reported as 236,643.34 tonnes (UNEP 1993). While 0.1 per cent of this is 237 tonnes, the fact that HCFCs have lower ODPs than CFCs would allow approximately 11–18 times as much to be produced. This gives an approximate value of 3,500 tonnes.

29 This figure is estimated by recognising that HCFC-22 and HCFC-141b/142b (the HCFCs that Atochem currently produce) were valued at US$2.50/kg and US$4.00/kg respectively (Greenpeace 1995: 16).

30 DGVI -- the directorate general for agriculture – was reported to be particularly keen to keep to the Copenhagen target (Reuters 1993).

31 'The argument for doing so is based above all on the risk of competition distortion with non-member Mediterranean countries' (AE 1993b). The proposed regulation was sent to the Parliament early in 1994. As with the HCFC proposal, the parliamentarians went further – in this case, they proposed a complete prohibition of the chemical by the end of 1999 (ENDS 1994a: 36).

Ministers, however, rejected this amendment in June, and the original proposal eventually came into effect.

32 Nevertheless, many still highlighted the potential economic impact of methyl bromide regulation. One report, for example, claimed that 'to ban the use of methyl bromide in Europe could lead to agricultural losses in France, Spain, Italy and Greece equivalent to PTS 300 bn/y. It is estimated that this could cost 14,300 jobs in Spain' (PE 1995: 51–52).

33 The US Environmental Protection Agency pushed for a global phaseout by 2001, for its officials thought that 'a level playing field would be most fair to those in the agricultural community, and the best and fastest way to develop alternative pest control tools' (USEPA, Methyl Bromide Home Page, WorldWideWeb).

34 Another report also notes that some countries feared 'the competitive advantage for States not bound by the control measures of the Protocol' (EPL 1996: 68).

35 I pursue this point after considering an 'interest-based analysis' of the climate change issue in Rowlands (1995b).

Bibliography

AE (1993a) 'Environment Council Agrees to Limit Substances Affecting the Ozone Layer,' *Agence Europe*, 4 December.

—— (1993b) 'Ministers Prepare Dossiers on Packaging and Waste,' *Agence Europe*, 7 October.

Benedick, Richard (1991) *Ozone Diplomacy: New Directions in Safeguarding the Planet* (London: Harvard University Press).

Chance, C. (1995) *The European Union – Understanding and Influencing Policy and Law-Making* (London: Clifford Chance).

DoE (1976) United Kingdom Department of the Environment, Central Unit on Environmental Pollution, *Chlorofluorocarbons and their Effect on Stratospheric Ozone*, Pollution Paper No. 5 (London: HMSO).

EB (1993) 'EC Weakens Montreal Protocol Targets,' *Environment Business*, 19 May.

EE (1995) *Europe Environment*, No. 462, 3 October.

ENDS (1992a) 'Ozone Layer Left at Risk by New Global Agreement,' *ENDS Report*, No. 214, November.

—— (1992b) 'HCFCs, Methyl Bromide to Dominate Ozone Layer Meeting,' *ENDS Report*, No. 213, October.

—— (1992c) 'Environment Ministers Agree Tougher Rules on CFCs,' *ENDS Report*, No. 215, December.

—— (1992d) 'Ministers Agree on Ozone Depleters, Sulphur in Gas Oil,' *ENDS Report*, No. 206, March.

—— (1993a) 'Brussels Launches New Talks on Ozone Depleters,' *ENDS Report*, No. 221, June.

—— (1993b) 'Majority Voting Looms on Landfills, Ozone Depleters,' *ENDS Report*, No. 225, October.

—— (1993c) 'Ministers Agree on Climate Treaty, Ozone Depleters, Packaging,' *ENDS Report*, No. 227, December.

—— (1993d) 'Brussels Prepares Controls on HCFCs, Methyl Bromide,' *ENDS Report*, No. 217, February.

—— (1994a) 'EC Institutions Battle Over Curbs on Ozone Depleters,' *ENDS Report*, No. 229, February.

—— (1994b) 'Go-Ahead for Controls on Ozone Depleters,' *ENDS Report*, No. 233, June.

—— (1995a) 'Ministers Face Deadlock at Ozone Layer Meeting,' *ENDS Report*, No. 248, September.

—— (1995b) 'Ministers Tighten EC Position on Ozone Depleters,' *ENDS Report*, No. 249, October.

—— (1995c) 'Ozone Layer Left at Risk as Talks Stumble on Funding,' *ENDS Report*, No. 251, December.

Engelmann, Frudolf (1982) 'A Look at Some Issues Before an Ozone Convention,' *Environmental Policy and Law*, Vol. 8, No. 2, pp. 49–56.

EPL (1996) 'Montreal Protocol: The Vienna Meeting,' *Environmental Policy and Law*, Vol. 26, Nos. 2/3.

Follett, Christopher (1992) 'France at Odds with EC Partners at Ozone Talks,' *Reuter News Service*, 24 November.

Gamlen, P. H. *et al.* (1986) 'The Production and Release to the Atmosphere of CCl_3F and CCl_2F_2 (Chlorofluorocarbons CFC-11 and CFC-12),' *Atmospheric Environment*, Vol. 20, No. 6, pp. 1077–1085.

Grant, Wyn, A. Martinelli and W. Paterson (1989) 'Large Firms as Political Actors: A Comparative Analysis of the Chemical Industry in Britain, Italy and West Germany,' *West European Politics*, Vol. 12, No. 2, pp. 72–90.

Grant, Wyn, W. Paterson and C. Whitson (1988) *Government and the Chemical Industry: A Comparative Study of Britain and West Germany* (Oxford: Clarendon Press).

Greenpeace (1995) *The Ozone Layer Destroyers: Whose Chlorine and Bromine Is It?* (Amsterdam: Greenpeace International).

Haas, Peter (1993) 'Stratospheric Ozone: Regime Formation in Stages,' in Oran R. Young and Gail Osherenko (eds), *Polar Politics: Creating International Environmental Regimes* (London: Cornell University Press).

Haigh, Nigel (1989) *EEC Environmental Policy and Britain*, second revised edn (Harlow: Longman).

Jachtenfuchs, Markus (1990) 'The European Community and the Protection of the Ozone Layer,' *Journal of Common Market Studies*, Vol. 28, No. 3, pp. 261–277.

Krueger, Jonathan and Ian H. Rowlands (1996) 'Institutions for Global Environmental Change: Protecting the Earth's Ozone Layer,' *Global Environmental Change*, Vol. 6, No. 3, pp. 245–247.

Layman, Patricia (1986) 'Aerosols Back on Road to Success,' *Chemical and Engineering News*, 28 April.

Litfin, Karen (1994) *Ozone Discourses: Science and Politics in Global Environmental Cooperation* (New York: Columbia University Press).

Love, Brian (1993) 'EC Plan for HCFC Phase-Out Stifled, France Blamed,' *Reuter News Service*, 18 March.

McConnell, Fiona (1991) 'Book Reviews,' *International Environmental Affairs*, Vol. 3, No. 4, pp. 318–320.

MacKenzie, Debora (1988) 'Now it Makes Business Sense to Save the Ozone Layer,' *New Scientist*, Vol. 120, 29 October.

MacNeill, J., P. Winsemius and T. Yakushiji (1991) *Beyond Interdependence: The Meshing of the World's Economy and the Earth's Ecology* (Oxford: Oxford University Press).

Maxwell, James and Sanford L. Weiner (1993) 'Green Consciousness or Dollar Diplomacy?: The British Response to the Threat of Ozone Depletion,' *International Environmental Affairs*, Vol. 5, No. 1, pp. 19–41.

Morrisette, Peter (1989) 'The Evolution of Policy Responses to Stratospheric Ozone Depletion,' *Natural Resources Journal*, Vol. 29, No. 3, pp. 793–820.

NS (1975) 'MP Calls for British Fluorocarbon Ban,' *New Scientist*, Vol. 67, 7 August.

—— (1978) 'Verdict Still Open on Fluorocarbons,' *New Scientist*, Vol. 79, 21 September.

OECD (1976) Organisation for Economic Co-operation and Development, *Fluorocarbons: An Assessment of Worldwide Production, Use and Environmental Issues (First Interim Report)* (Paris: OECD).

Oye, Kenneth and James Maxwell (1995) 'Self-Interest and Environmental Management,' in Robert O. Keohane and Elinor Ostrom (eds), *Local Commons and Global Interdependence* (London: Sage Publications).

Parson, Edward (1993) 'Protecting the Ozone Layer,' in Peter Haas, Robert Keohane and Marc Levy (eds), *Institutions for the Earth: Sources of Effective International Environmental Protection* (Cambridge, Mass.: MIT Press).

Parson, Edward and Owen Greene (1995) 'The Complex Chemistry of the International Ozone Agreements,' *Environment*, Vol. 37, No. 2, March, pp. 16–20, 34–43.

PE (1995) *Phytoma España*, No. 69, 31 May.

Purvis, M. (1994) 'Yesterday in Parliament: British Politicians and Debate Over Stratospheric Ozone Depletion, 1970–92,' *Environment and Planning C: Government and Policy*, Vol. 12, No. 3, pp. 366–367.

Reuters (1992) 'EC Ministers Attack New Ozone-Eating Chemical,' *Reuter News Service*, 20 October.

—— (1993) 'EC Proposals on HCFCs and Methyl Bromide Due on May 12,' *Reuter News Service*, 3 May.

Rowlands, Ian (1993) 'The Fourth Meeting of the Parties to the Montreal Protocol: Report and Reflection,' *Environment*, Vol. 35, No. 6, pp. 25–34.

—— (1995a) *The Politics of Global Atmospheric Change* (Manchester: Manchester University Press).

—— (1995b) 'Explaining National Climate Change Policies,' *Global Environmental Change*, Vol. 5, No. 2.

Schoon, Nicholas (1990) 'Deal to Save Ozone Layer Agreed,' *The Independent*, 30 June.

Thornton, Grant (1990) *1989 Production and Sales of Chlorofluorocarbons 11 & 12* (Washington, DC: Chemical Manufacturers Association, Fluorocarbon Programme Panel).

UKSORG (1988) United Kingdom Stratospheric Ozone Review Group, *Stratospheric Ozone* (London: HMSO).

UNEP (1992) UNEP/OzL.Pro.4/2.

—— (1993) 'The Reporting of Data by the Parties to the Montreal Protocol on Substances that Deplete the Ozone Layer,' UNEP/OzL.Pro.5/5, 24 August.

Van Haasteren, J. (1994) 'Methyl Bromide Damages the Ozone Layer,' *Change*, No. 18, p.16.

WCED (1987) *Our Common Future* (Oxford: Oxford University Press).

3

REGULATING EXPORTS OF HAZARDOUS CHEMICALS

The EU's external chemical safety policy

Marc Pallemaerts

Over the past decade, the EU has succeeded in developing an external chemical safety policy and in asserting its role as a major actor in the global debate on trade in banned and severely restricted chemicals. This chapter analyses the development of this policy, and its dependence on the internal dynamics of EU law-making as well as upon EU relations with third countries.

The issue of international trade in banned and severely restricted chemicals emerged on the international environmental policy agenda in the early 1980s, as a result of concern expressed by Third World governments in United Nations fora (Pallemaerts 1988) and reports published by various non-governmental organisations (Weir and Schapiro 1981, Bull 1982) about the serious health and environmental effects of the export to developing countries of hazardous chemicals, especially pesticides, whose use is prohibited or subject to severe regulatory restrictions in producing countries. As the world's largest exporter of pesticides,[1] the EC found itself under pressure from the international community and from a broad coalition of environmental, consumer and development NGOs to control trade in banned and severely restricted products.[2] The European Commission played a key role as a focal point for political pressure, as policy initiator in the internal legislative process, and later also as spokesperson in external policy negotiations. In developing an external chemical safety policy, it had to walk a tightrope between global health and environmental concerns on the one hand and the European chemical industry's trade interests and competitiveness concerns on the other.

The slow-motion development of an EU policy on hazardous chemical exports

European Community chemical safety legislation was originally concerned exclusively with the internal market. The exclusion of chemicals intended for export from the scope of such Community legislation was no more than a logical consequence of a regulatory policy focused on the internal market and was not perceived as a political issue until the early 1980s.[3] The legal vacuum with respect to exports was obviously in harmony with the interests of the EC chemical industry, heavily dependent as it is on export markets. In 1978, EC-based companies accounted for 38.6 per cent of world pesticide sales, and the five largest of them (Bayer, Shell, ICI, Rhône-Poulenc and BASF) made 65 per cent of their total sales outside Europe (Chetley 1985: 32). In 1981, 31 per cent of total EC exports of pesticides went to developing countries. The main pesticide-exporting member states were Germany, the United Kingdom, France, the Netherlands and Italy, and their respective shares of overall EC pesticide exports to the Third World in the same year 32 per cent, 24 per cent, 18 per cent, 10 per cent and 5 per cent.[4]

As a result of this configuration of interests, the development of an external Community policy on chemical safety was very slow. The Commission initially did not regard such a policy as necessary. The earliest political initiatives in this field were actually taken by the European Parliament in the face of strong Commission opposition. It was only in the mid-1980s that the Commission, confronted not only with political pressure from the Parliament and from NGOs, but also from a number of member states mental organisations such as the OECD, UNEP and FAO, changed its position and eventually submitted a proposal for legislation on the export of certain hazardous chemicals to the Council in 1986.[5]

Prior to this legislative initiative, the Commission consistently opposed regulation of hazardous chemical exports on the grounds that 'it should be for the importing country to lay down its own rules for trade in these products' (EP 1982a) and that these rules 'may differ from Community provisions for sound and objective reasons' (EP 1982b). Yet the Commission's policy was not entirely without contradictions. Ironically, when questioned in 1981 about possible exports of banned hazardous consumer products from the United States to the European Community, the Commission asserted 'that every country engaged in international trade must ensure that any products liable to be a direct hazard to users or consumers should not be exported' (EP 1981a). However, at that time it was clearly not prepared to apply the same principle to the Community's own exports of toxic chemicals. To rationalise this double standard, the Commission relied on the familiar argument that local conditions in developing countries may lead to a different risk–benefit assessment, and that any attempt to 'impose Community rules on non-member countries or deny them supplies of pesticides they need could cause vexed political problems' (EP 1982a).

Later, however, the Commission, while still unwilling to take any action to amend Community directives, expressed some support for the concept of organised information exchange on hazardous chemicals in international trade, but deferred to the initiative of other international organisations in this field (EP 1982a). In particular, the Commission seemed to favour the FAO as the most appropriate forum to address the issue, and it participated in the consultations on the FAO International Code of Conduct on the Distribution and Use of Pesticides (EP 1982b, EP 1983b: 240). In its statements about the desirability of international notification arrangements for hazardous products generally, the Commission continued to place heavy emphasis on the 'observance of [the] sovereign rights of each state', stressing 'that it is up to each country to decide whether a product may be imported or not' (EP 1981b: 16).

In 1983, in response to a motion for a resolution introduced by the Socialist Group, the European Parliament's Environment Committee adopted a report recommending that exports of banned pesticides from the EEC be made subject to notification and the express consent of the importing country and that all exported pesticides should comply with EEC packaging and labelling standards (EP 1983a). On 14 October 1983, the European Parliament adopted the resolution proposed by its Environment Committee. This resolution recommended amendments to existing Community pesticides control legislation to make pesticide exports subject to the twin conditions

(a) *that the government of the importing country is informed* of the particular nature of the product and of the restrictions to which it is subject in the exporting country and the reasons for such restrictions;
(b) *that the government of the importing country*, having received such notification, *explicitly requests the purchase.*

(EP 1983c, emphasis added)

Additionally, the resolution called for legislation to ensure that all pesticides exported from the Community to developing countries are packaged and labelled in accordance with the standards laid down for their placing on the market within the Community and 'that the directions for use are written in the most common language of the country of destination, preferably accompanied by diagrams, unless specified otherwise by the importing countries'.

In 1983, the European Parliament resolution was the clearest pronouncement thus far by a political institution from the pesticide-exporting countries in favour of a system of export controls to ensure the prior informed consent (PIC) of the importing country. However, the resolution had no immediate effect on the position of the Commission, which continued to oppose any export regulation by the Community.

During the plenary debate in Parliament, the European Commissioner then in charge of environmental affairs, Karl-Heinz Narjes, criticised the resolution for seeking to force EC standards on other countries, and reiterated the Commission's standard position that 'the autonomy and the principle of respect of the sovereign rights of these third countries would thereby be affected, [while] these countries must and can make their own decisions concerning the safety and health of their citizens' (EP 1983b).

Towards an EEC regulation on exports of banned and severely restricted chemicals

Several factors contributed to prompting a change in the Commission's negative attitude towards regulation of international trade in banned and severely restricted chemicals. Apart from the strong position taken by the European Parliament in its resolution of 14 October 1983, there were also indications of support for Community action on the part of some member states. In June 1983, the Dutch government, in an attempt to prod the Commission into action, put the pesticides export issue on the Council agenda and submitted a memorandum suggesting 'that Community rules be adopted regarding the export of specified pesticides to countries outside the Community' (ND 1983). In response to this proposal, the Council requested the Commission to prepare a report on the issue and referred the matter to the Committee of Permanent Representatives for further discussion (TSCN 1983: 2).

Apart from the Council request and the European Parliament resolution, other factors which seem to have incited the Commission to act are the fact that various intergovernmental organisations such as the OECD, UNEP and FAO were developing regulatory instruments and that several member states were in the process of enacting national legislation on exports of chemicals. In 1985, the Netherlands and the United Kingdom, followed by the Federal Republic of Germany in 1986, introduced new legislation providing statutory authority for certain measures to regulate exports of banned or severely restricted chemicals. The taking of unilateral legislative action by a number of member states, all three of them major exporters of chemicals, threatened to cause trade distortions and was a strong incentive for Community action.

The Commission took a long time to elaborate concrete proposals for Community action, still fearing that unilateral action would jeopardise EC trade interests. Initially, it merely observed developments in international fora and waited for the outcome of the ongoing negotiations on export notification schemes in UNEP and FAO before taking any initiative of its own. The Commission participated in the UNEP and FAO negotiations but acted as a follower rather than a leader in the international regulatory debate. Its position in these negotiations was essentially conservative, as it joined the majority of EC member states and other OECD countries in an effort to ensure that the non-binding international standards for the responsibility of exporting countries being established by UNEP and FAO did not go beyond mere information exchange.

Although it initially opposed the principle of PIC in FAO and UNEP, the Commission eventually changed its position on this question too and moved closer to the Parliament's view that regulation of exports of banned and severely restricted chemicals must be based on the PIC principle. This important shift in the Commission's position was prompted by discussions in the OECD on the control of exports of hazardous wastes. In March 1985, an OECD ministerial conference adopted a recommendation urging member states not to allow the export of hazardous wastes, in particular to developing countries, unless the importing country possesses 'adequate disposal facilities' and has given its formal consent (OECD 1985). Thus, the member states of the OECD, which in 1984 explicitly rejected export controls with respect to banned and severely restricted chemicals (Pallemaerts 1988: 65), agreed in 1985 that hazardous wastes should not be shipped abroad without the consent of the importing country. In June 1986, this OECD recommendation was translated into Community law through an amendment of the 1984 EEC Directive on transfrontier shipments of hazardous wastes (ECM 1986).

Barely two weeks after the adoption of this amendment by the Council, the Commission officially put before the Council its proposal for a regulation concerning exports of dangerous chemicals (EC 1986b), in which it proposed to make such exports subject to the principle of 'prior informed choice'. In explaining this proposal, the Commission described the issue of hazardous exports as 'one of the most important political issues of the moment in respect to environmental matters and international trade' (EC 1986a: 4) and established a clear parallelism between the export of banned and severely restricted chemicals and that of hazardous waste, by referring in its explanatory memorandum to the 1985 OECD recommendation.[6]

The Commission's 1986 'informed choice' proposal

The Commission's original proposal (EC 1986b) for what eventually became Regulation (EEC) No. 1734/88, is worth analysing in some detail, in view of the originality of the informed choice system proposed by the Commission. Even though this system was not endorsed by the Council, the proposal nevertheless constituted an important political decision of the Commission of the European Communities and a significant contribution to the international regulatory debate in its own right.

After its initial reluctance to get involved in the regulation of chemical exports, the Commission realised that the Community as a major trading power could not ignore this issue and that the common commercial policy was in fact the most appropriate framework for addressing it. At the time it was formulated, the proposed regulation was the most far-reaching and interventionist international regulatory proposal, which contrasted sharply with the minimalist approach which had initially prevailed in other international fora. The measures proposed by the Commission were more stringent than anything contained in the then existing recommendatory instruments developed by OECD, UNEP and FAO (Pallemaerts 1988). It was also especially significant as the first proposal for a *binding* legal instrument to regulate exports of banned and severely restricted chemicals.

The Commission proposed to introduce restrictions on the export of dangerous chemicals in two stages. In a first stage, it aimed to 'bring Community practice into line with existing international codes', by establishing a mandatory export notification procedure, which would ensure the application of those existing international standards 'at Community level and in a uniform manner' (EC 1986a: 2).[7] The informed choice system, as proposed by the Commission, would only have come into effect two years after the entry into force of the regulation. The Commission hoped that, in the meantime, UNEP and OECD could be convinced to include the same system in their respective instruments, and it requested a mandate from the Council to negotiate on behalf of the Community in those organisations.

Under the proposed informed choice system, the export from the Community of the chemicals covered by the regulation would have required an authorisation to be issued by the designated authority of the member state of exportation. This export authorisation would have been issued not only 'if the country of destination consents to the import of the chemical concerned', but also '*if no communication is received from that country within 60 days* of the date on which notification was sent by the Commission'. The system proposed by the Commission was a somewhat imperfect implementation of the principle of prior informed consent. The *explicit* consent of the importing country, which was called for in the 1983 European Parliament resolution, would not actually have been required. If the importing country did not react within sixty days, its silence would have been interpreted as *tacit* consent. This is apparently why the Commission opted for the term 'informed choice' instead of the more explicit term 'prior informed consent'.

The Commission's proposal also comprised a recommendation for a Council decision authorising the Commission to negotiate on behalf of the Community within the framework of the OECD and UNEP. Having at last endorsed some form of informed consent, the Commission proposed to take the lead in the international regulatory debate and to try and 'export' the informed choice formula to other international fora. The Commission's objective in these fora was to seek to ensure that exporters in other countries would be subject to similar constraints as those to be imposed on EEC exporters.

Community internal legislation

The 1988 Export Notification Regulation

When finally adopted by the Council in June 1988, the regulation (ECM 1988a, hereafter referred to as Regulation 1734/88) was substantially weaker than the Commission's initial proposal.

The regulation was based on Article 130s of the Treaty, instead of Article 113 as originally proposed by the Commission.[8] The Council did not follow the Commission in considering the regulation as a trade policy measure, but chose to treat it as an environmental policy measure under Article 130s, in view of its objective of protecting health and the environment. This choice was probably based primarily on political considerations related to the decision-making procedure. Apparently the member states were reluctant to accept the decision-making by qualified majority which would have resulted from the choice of Article 113 EEC as legal basis, and opted instead for Article 130s because, at that time, it provided for unanimity.[9] The Commission had argued strongly that Article 113 was the proper legal basis for this measure, and entered a formal statement in the minutes of the Council criticising the Council's decision to change the legal basis to Article 130s, and reserving the right to challenge that decision in Court (AE 1987). It did not, however, carry out this threat and initiate annulment proceedings against the regulation.

The Council did not accept the informed choice procedure proposed by the Commission, but opted instead for a minimal notification system modelled on the OECD Guiding Principles and corresponding FAO and UNEP provisions. The Council rejected the Commission's view that the Community should take the lead internationally by applying PIC unilaterally, while simultaneously trying to convince other international fora to do the same. The Council did not wish to risk imposing on Community exporters a competitive disadvantage but chose to await further developments at the international level before taking any initiatives in this field.

Together with the regulation, the Council adopted a resolution which, referring implicitly to the Dutch initiative to introduce PIC at the national level, welcomed 'action by Member States to test the practical value' of PIC and invited the Commission

> to examine this question in greater detail and to submit, where necessary, in the light of the information supplied by the Member States and *developments in relevant international practices*, appropriate proposals with a view to the possible adjustment of the Community instrument.
>
> (ECM 1988b)

This resolution, which also noted the Commission's intention to participate in the UNEP negotiations on PIC, clearly indicated that developments in other international fora would be a key determinant of further action by the Community. Thus the Community elected to remain a follower rather than become a leader.

Regulation 1734/88 established 'a common system of notification and information' (ECM 1988a: art. 1). This system applies to a common list of chemicals banned or severely restricted *by Community law*, as laid down in Annex I of the regulation. It provides for a single export notification for the Community as a whole. This single notification covers 'every subsequent export of the chemical concerned from the Community to the same third country' (ECM 1988a: art. 4, para. 3). There is no notification requirement for chemicals which are banned or severely restricted in individual member states.

The provisions of the regulation left the member states considerable discretion as to *how* notification will be effected. It only provided that the designated authority of the exporting member state 'shall take the necessary measures to ensure that the appropriate authorities of the country of destination receive notification' (ECM 1988a: art. 4, para. 1). This language does not explicitly require that the designated authority *itself* give notification. It was possible for a member state to delegate that responsibility to the exporter, provided it could ensure that the notification requirement is effectively complied with. As a matter of fact, this eventuality was explicitly envisaged in a footnote to Annex II of the regulation, which specified the information to be given 'where the designated authority provides that the notification be made *by the exporter*'.

According to a 1993 Commission report on the implementation of Regulation 1734/88, in a period of over three years, from the entry into force of the regulation until 1 November 1992, not more than sixty export notifications were effected pursuant to it for the Community as a whole. Only six member states reported notifications to the Commission. The other six confirmed to the Commission that 'to their knowledge' they did not export any of the products covered by the regulation (EC 1993b: 3). Of the twenty-one substances or groups of substances listed in Annex I to the regulation, only eight were the subject of notifications. In all, notifications were addressed to only some thirty importing countries (EC 1993a: annex A2). Whether due to non-compliance or to other causes, the small number of notifications, products and third countries of destination involved indicates the limited significance of Regulation 1734/88. It does not seem to have imposed any real constraints on trade. The measure appears to have been largely symbolic, in that most of the products which were made subject to the notification requirement were no longer produced in the Community or traded on any significant scale in the first place.

The 1992 Prior Informed Consent Regulation

In a compromise between advocates and opponents of PIC among the member states, the Council, when it adopted Regulation 1734/88, had undertaken to reconsider the introduction of a PIC procedure at Community level two years later and invited the Commission to submit appropriate proposals before July 1990. The Commission, in a formal statement recorded in the Council minutes, had expressly 'dissociated itself' from the Council decision not to translate the principle of PIC into EC law immediately, as it had originally proposed in 1986. The Commission had reaffirmed its commitment to PIC as 'the only acceptable basis for the Community's exports of chemical products whose use is banned or strictly limited in the Community' and announced its intention to ask the Council to reconsider its rejection of the principle, noting 'growing international pressure for its adoption' (AE 1987). One could say that there was in fact a political understanding between the Council and the Commission that the matter would be reconsidered on the basis of the outcome of the negotiations on PIC that were then going on in UNEP and FAO.

The Commission closely followed those negotiations and, as soon as agreement on a PIC scheme started taking shape in those international fora, began to draft proposals for amending Regulation 1734/88. It is quite symptomatic of the recognised inadequacy of that regulation that proposals for amending it were already being drafted even before it had entered into force on 22 June 1989.[10] Nevertheless, the Commission did not manage to meet the target date of 1 July 1990 mentioned in the preamble of the 1988 regulation, but submitted its proposal to the Council a few months later, on 17 December 1990 (EC 1991). The new regulation concerning Community exports and imports of certain dangerous chemicals was eventually adopted by the Council on 23 July 1992 (ECM 1992). It entered into force on 29 November 1992, replacing Regulation 1734/88, which it formally repealed (ECM 1992: art. 12, para. 1).

The 1992 regulation expands considerably on the 1988 regulation, by supplementing the Community export notification system, in force since 1989, with a legally binding PIC procedure, based on the voluntary scheme contained in the UNEP London Guidelines and the FAO Code of Conduct, both as amended in 1989. Apart from introducing the PIC procedure, Regulation 2455/92 also makes a number of improvements in the existing export notification procedure. It specifies that the designated national authority of the member state of export shall *itself* notify the country of destination, and that the exporter is under an obligation to provide it with the information necessary to that effect no later than thirty days before export is due to take place. The new regulation also provides that the export notification 'shall as far as possible take place at least 15 days before export' (ECM 1992: art. 4, para. 1).

The export notification procedure, originally introduced by the 1988 regulation, continues to apply only to chemicals banned or severely restricted under Community law. The list of these chemicals, laid down in Annex I of Regulation 2455/92, is an expanded version of Annex I of Regulation 1734/88.[11] The 1992 regulation also strengthens the packaging and labelling requirements for exported chemicals, initially introduced by Article 5 of Regulation 1734/88. These requirements now apply not only to chemicals banned or severely restricted under Community law, as provided in the 1988 regulation, but generally to all 'dangerous chemicals . . . intended for export' which would be subject to Community provisions relating to packaging and labelling if they were marketed within the EC (ECM 1992: art. 7, para. 1).[12] However, a provision specifying that the requirement of compliance with EEC packaging and labelling standards 'shall be without prejudice to any specific requirements of the importing country' (ECM 1992: art. 7, para. 1) emphasises the priority to be given to the standards of the importing country and stresses the complementary function of the EEC standards.

The most important provisions of Regulation 2455/92 are those relating to the PIC procedure. The necessary information for the application of the PIC procedure in the Community is to be incorporated in Annex II of the Regulation. According to Article 5, para. 3, this Annex shall include the international list of banned and severely restricted chemicals subject to the PIC procedure, as established jointly by UNEP and FAO, the list of countries participating in the UNEP/FAO PIC scheme and the decisions taken by these countries regarding the import of the listed chemicals. The purpose of including all this information in an annex to the regulation is to make the international PIC procedure enforceable as part of European Community law. Indeed, the regulation explicitly provides that 'the exporter shall be required to comply with the decision of the country of destination participating in the PIC procedure' (art. 5, para. 4). However, the enforcement measures to be taken are left to the discretion of the member states, as Article 6 only provides that they 'shall take appropriate legal or administrative action in case of infringement'. The Commission had initially proposed the imposition of 'severe or dissuasive sanctions . . . on persons exporting chemicals subject to the international PIC procedure . . . contrary to the PIC decision of the country of destination' (EC 1991: art. 5, para. 4), but that requirement apparently proved too specific for certain member states.

It should be noted, however, that when the regulation was formally adopted in July 1992, Annex II was left blank. In the text of Regulation 2455/92 as published in the *Official Journal*, Annex II bears the heading 'Chemicals subject to the international PIC procedure and the PIC decisions of importing countries', but contains no entries. The first entries in Annex II were formally adopted in January 1994 (EC 1994).[13] Thus, it was only one and a half years after the adoption of Regulation 2455/92 that the Community actually took the necessary legislative action to make the PIC provisions of the regulation operational. Ironically, in May 1993, the Commission published a brochure on the Community's policy with respect to exports of banned or severely restricted chemicals (EC 1993a), boasting that Regulation 2455/92 'makes the PIC procedure mandatory for the export of chemicals from the Community to participating countries' and 'could serve as a model for a workable international PIC system'. However, at the time of the brochure's publication, the Community's much-hailed mandatory PIC procedure was not yet operational, for lack of any entries in Annex II of the regulation.

The delay in making the PIC provisions operational is due to the formal procedure laid down in the regulation for any amendments to Annex II. This procedure requires the Commission to submit the proposed amendments to a committee composed of member state representatives, which shall give an opinion by qualified majority vote (ECM 1992: art. 11, para. 2). Some member states have used the committee procedure to delay the adoption of Annex II and its subsequent modifications. Their attitude is contrary to the spirit and letter of the UNEP/FAO voluntary PIC scheme, under which changes to the list of PIC chemicals and PIC decisions of importing countries are not subject to the discretion of EC institutions but result automatically from decisions taken by an international expert committee or by importing countries and notified by UNEP/FAO. Exporting countries have no discretion whether or not to 'adopt' these changes but are committed to apply the PIC list as established by UNEP/FAO and to respect the decisions of importing countries. Amendments to Annex II should in fact be a mere formality, but the procedure laid down in the regulation makes it impossible for the Community to keep up with the PIC decisions which are communicated by UNEP/FAO on a routine basis every six months.

The implementation and effectiveness of Regulation 2455/92

In July 1995, the European Commission, in cooperation with the European Parliament, organised a three-day 'Conference on International Trade in Dangerous Chemicals' in Brussels. This conference, which had no formal decision-making authority, was intended to assess the effectiveness of Regulation 2455/92, to analyse the chemicals management situation in developing countries and the role of information exchange and PIC procedures in improving it, and to formulate recommendations for future measures and policy in this field. The conference was attended by representatives of all interested parties, including government representatives from EU member states and non-EU countries, in particular chemical-importing developing countries, as well as representatives of various intergovernmental organisations with an interest in chemical safety, industry, trade unions and NGOs.

In preparation for the conference, the Commission had commissioned the

German aid agency GTZ to conduct a survey of designated national authorities in importing countries and other interested actors to evaluate the impact of the EU's export notification and PIC procedures. This survey showed that many authorities in importing third countries were not familiar with the EU measures of which they are the supposed primary beneficiaries, that notifications often did not reach the competent authorities and that the information provided in export notifications was inadequate to meet the needs of importing countries.

During the conference it was stressed by all participants that information exchange and PIC procedures were not in themselves sufficient to ensure chemical safety in developing countries and that more efforts should be devoted to training and capacity building. Several suggestions were formulated by participants for improving the existing EU measures: providing more information in export notifications, especially information concerning the port of entry and identity of the importer, developing provisions for more effective control of shipments in cooperation with customs authorities, extending export notification requirements to all chemicals subject to PIC, including those which are not banned or severely restricted by the EU, etc. Participants also called on the EU to contribute actively to the development of a global, legally binding instrument on PIC.

In his closing statement, the representative of the Commission agreed that a revision of Regulation 2455/92 was necessary and announced that a proposal to this effect would be prepared as soon as possible. Indeed, amendments to Annex III of the regulation, which lists the information to be provided to importing countries in export notifications, were adopted in July 1996 (EC 1996). More information must now be supplied in the notifications, including the identity of the importer. It seems rather unlikely, however, that any further revisions of the regulation will be proposed before the conclusion of the UNEP/FAO negotiation process on a PIC convention.[14]

The role of the European Community in multilateral negotiations

GATT/WTO negotiations on 'domestically prohibited goods'

The issue of international trade in banned and severely restricted products was first raised within the forum of the GATT by Nigeria and Sri Lanka in 1982. As a result of this initiative, the 38th session of the Contracting Parties adopted a decision on 'export of domestically prohibited goods' providing that 'contracting parties shall, to the maximum extent feasible, notify GATT of any goods produced and exported by them but banned by their national authorities for sale on their domestic markets on grounds of human health and safety' (BISD 1983: 19). This decision, taken before the adoption by the OECD, UNEP and FAO of their specific recommendations on notification procedures for hazardous chemicals, was in fact the first international instrument requiring governments to provide information on banned products in international trade, but it should be noted that it was different in scope from those later recommendations, in that it applied only to products *banned* on *health or safety* grounds, and thus did not cover severely restricted products nor those banned for *environmental* reasons.

At the meeting of the GATT Trade Negotiations Committee at ministerial level, held in Montreal in December 1988, in the wake of the public outrage caused in the Third World by the uncovering of a series of projects to dump hazardous waste from Europe and North America in debt-ridden African countries, a group of developing countries headed by Nigeria and Cameroon undertook a renewed effort to put the issue of trade in hazardous substances, including wastes, on the agenda of the Uruguay Round (GATT 1989: 4–5). This initiative encountered fierce opposition from industrialised countries, who considered that the problem was already being adequately addressed in the competent 'technical' fora like UNEP, the FAO and OECD, and that GATT negotiations on the subject would duplicate the ongoing efforts of those organisations. Due to this opposition, the issue was not added to the Uruguay Round agenda, but, as a compromise, developed and developing GATT contracting parties agreed to step up consideration of the problem under GATT's regular work programme, and more specifically to discuss '*complementary action that may be necessary in GATT*', having regard to the work that is being done by other international organisations' (GATT 1988a, emphasis added).

Six months later, the GATT Council formally established a 'Working Group on the Export of Domestically Prohibited Goods and Other Hazardous Substances', which was instructed to 'examine trade-related aspects that may not be adequately addressed' in other international agencies' work in this field 'in the light of GATT obligations and principles'. In its decision, the Council again stressed 'the need to avoid duplicating the work of other international organisations' (BISD 1990: 402–403).[15]

It is in the GATT Working Group on Domestically Prohibited Goods that the European Commission, negotiating on behalf of the European Community and its member states, for the first time assumed an active role in the international regulatory debate on trade in hazardous products. As a matter of fact, it was forced to do so because of its institutional position within the GATT and the need for the Community to speak with a single voice in that forum.

Much of the negotiating effort in the Working Group was devoted to accommodating the parties' conflicting views as to the need for the development of general rules on hazardous exports within the framework of GATT as a complement to the special rules and procedures for specific categories of hazardous products elaborated by the different technical organisations. The key question was the determination of the proper scope of application of the proposed GATT rules and their relation to the existing sectoral schemes. In the view of the African and Asian contracting parties who put the issue on the GATT agenda, the rules to be developed within the GATT, providing for a general export ban or at least an export licensing system, were intended to override the less restrictive information exchange or even PIC procedures recommended by the specialised organisations. The industrialised countries, however, sought to preserve those procedures, which they considered quite adequate, and did not wish to see them superseded by more restrictive GATT rules.

The EC tried to bridge the gap between the industrialised and developing countries' positions by submitting to the Working Group, in early 1990, a draft 'Understanding on Trade in Domestically Prohibited Goods and other Hazardous Substances' (GATT 1990). In this draft, in what now sounds almost like a direct rebuttal of the *Tuna–Dolphin* panel's interpretation of GATT art. XX a few years later, it explicitly

> recognise[d] the need for governments, in formulating policy in respect of trade in goods . . . to pay the fullest attention possible to the protection of the environment, and of human, animal and plant health and life, *not only within their own countries but also in other countries*.
>
> (GATT 1990: art. 2, para. 1)

This provision of the EC proposal, in an apparent attempt to find some common ground between North and South, echoed similar clauses in the proposals tabled by African and Asian countries (GATT 1988b, 1988c: 3, para. ii) and inspired compromise wording in a draft decision elaborated by the Chairman of the Working Group, which served as a basis for further negotiations.

However, the Working Group was unable to reach consensus on this draft decision. At the time of expiry of its extended mandate, the group had managed to reach consensus on the wording of a draft decision among all parties except the United States (GATT 1991), which continued to have reservations both as to the legal form and the content of the proposed instrument.[16] Due to this US opposition, the group was unable to complete its work successfully and the proposed decision was never formally adopted.

At the conclusion of the Uruguay Round in 1994, the issue of exports of domestically prohibited goods was included in the terms of reference of the newly established GATT/WTO Committee on Trade and Environment (GATT 1994). However, it is the last of the seven items on the CTE's agenda and one which was not given priority in the Committee's work programme in preparation for the WTO ministerial conference in Singapore in December 1996. In the CTE, Nigeria reintroduced a draft decision on domestically prohibited goods (WTO 1996a), based on the GATT Working

Group's unsuccessful 1991 draft. The EU, however, did not reiterate its 1990 proposal. The discussion in the CTE, largely a rehash of the earlier debate in the GATT Working Group, has so far remained inconclusive. In its report to the Singapore meeting, the CTE merely noted that it 'needs to continue to concentrate on *what* contribution *could* be made in this area by the WTO, bearing in mind the need for this work neither to duplicate nor to deflect attention from the work of other specialised inter-governmental fora' (WTO 1996b: para. 202, emphasis added) and, 'in the meantime' recommended further information-gathering efforts by the WTO Secretariat in cooperation with member states (WTO 1996b: para. 203).

UNEP/FAO Intergovernmental Negotiating Committee

The prime locus of multilateral negotiations on the regulation of international trade in banned or severely restricted chemicals has again shifted to UNEP/FAO, as a result of the decision of the UNEP Governing Council in May 1995 to launch formal intergovernmental negotiations, in cooperation with FAO, for the purpose of elaborating an international legally binding instrument for the application of the PIC procedure. The first two meetings of the 'Intergovernmental Negotiating Committee for an International Legally Binding Instrument for the Application of the Prior Informed Consent Procedure for Certain Hazardous Chemicals in International Trade' (INC/PIC), convened jointly by UNEP and FAO, were held respectively in Brussels and in Nairobi in March and September 1996. During these meetings, the EU has effectively asserted itself as one of the chief actors in those negotiations, the Commission having obtained a mandate from the Council to negotiate on behalf of the Community and its member states, given the existence of internal EU legislation on PIC.

The decision by the UNEP Governing Council to start negotiations for the conclusion of a legally binding multilateral instrument on PIC was the result of an excruciatingly slow consensus-building process which took six years and involved a variety of international fora: UNEP, FAO, UNCED, the UN Commission on Sustainable Development (CSD) and, most recently, the Intergovernmental Forum on Chemical Safety (IFCS) (Pallemaerts 1990–1994). Indeed, the 'possible further need for a convention' in this field was first mentioned in a 1989 decision of the UNEP Governing Council,[17] but it proved impossible to reach agreement on actually starting negotiations before 1995.

While the INC/PIC was expressly given 'a mandate to prepare an international legally binding instrument *for the application of the prior informed consent procedure* for certain hazardous chemicals in international trade' (UNEP 1995a, emphasis added), the exact scope of the instrument to be negotiated remains controversial. An apparent consensus, which had emerged from the deliberations of a UNEP *ad hoc* working group of experts and task force in

1993 and 1994, to limit the scope of the future legally binding instrument to making the application of the PIC procedure mandatory, as currently laid down in the voluntary UNEP and FAO instruments (Pallemaerts 1993, 1994), was shattered during the 1994 meeting of the CSD, where, as the Chairman's summary of the high-level segment of that meeting notes, a 'strong sentiment was expressed by participants' that the proposed instrument should not only make the PIC procedure legally binding, but 'subsequently ban the export of domestically prohibited chemicals from countries that are members of the Organisation for Economic Cooperation and Development to other countries' (UN 1994: 53, para. 11, emphasis added), by analogy with the 1994 decision of the Basel Convention contracting parties 'to prohibit immediately all transboundary movements of hazardous wastes which are destined for final disposal from OECD to non-OECD States' (MoP 1994).

The same demand was forcefully reiterated at the 1995 Governing Council meeting by the Group of 77, led by Malaysia, which tabled an amendment to the draft Governing Council decision stipulating that the proposed international instrument should include provisions banning 'the export of domestically prohibited chemicals, including pesticides' (UNEP 1995a).[18] In the ensuing negotiations, a compromise was finally reached on a 'double-track' approach: while the mandate of the INC remained limited to the application of the PIC procedure, it was agreed that, in parallel with the activities of the committee, a group of experts would be convened, also in cooperation with FAO, to 'consider ... and recommend what *further measures* are needed to reduce the risks from a limited number of hazardous chemicals, *either within or beyond the scope of the existing prior informed consent procedure*' (UNEP 1995c, emphasis added). This group was instructed to report to the nineteenth session of the Governing Council, which would 'give consideration to the need to develop further measures ... including the possibility that the mandate of the intergovernmental negotiating committee ... *be extended to provide a basis for such measures*' (UNEP 1995c, emphasis added).

Argument about the desirable scope of the PIC instrument mobilised most of the negotiators' attention during the first meeting of the INC. However, the debate took on a somewhat different form from what might have been expected, based on the earlier discussions in the CSD and the UNEP Governing Council. There was no coordinated G77 position and none of the countries that had supported the call for an export ban on previous occasions argued that the legally binding instrument should contain provisions banning the export of domestically prohibited chemicals from OECD to non-OECD countries. Instead, the debate focused on the issue of whether or not the instrument under negotiation should be a framework convention broader in scope than the current voluntary PIC scheme as laid down in the London Guidelines and FAO Code of Conduct, and, more specifically, on the desirability of including in the instrument provisions on export notification for banned or severely restricted chemicals, and a mechanism for adopting further control measures, including possible provisions for eventually phasing out or banning the *production*, rather than the export, of certain hazardous chemicals.

The call for extending the scope of the instrument beyond PIC this time did not come from the G77, but from European countries. In his opening speech at the Brussels meeting, the Belgian environment minister suggested that the legally binding instrument, in addition to the actual PIC provisions, should also provide a legal framework for the adoption of more drastic international action with respect to certain hazardous chemicals, including global production phaseout measures, as and when an international consensus on such measures emerged. He referred specifically to the measures currently under consideration in UNEP and the IFCS for phasing out certain persistent organic pollutants (POPs) and proposed that the instrument under negotiation in the INC/PIC should provide a legal basis for such measures (UNEP 1996a: 5).

While not being quite as specific as to the nature of possible further measures, the EU, during the negotiations, took the position that the scope of the instrument should not be strictly limited to the PIC procedure but should be flexible enough to be able to accommodate future developments. It proposed that the legally binding PIC instrument, to this end, should contain a framework provision providing a basis for the negotiation and adoption of additional protocols to the instrument at a later stage. This view was strongly opposed by the United States, Australia, Canada and some other OECD countries, who argued that the INC should strictly stick to its mandate and not attempt to elaborate any provisions going beyond the application of the PIC procedure.

At the experts meeting on 'further measures', which was held in Copenhagen in April 1996, the Netherlands and Belgium jointly introduced a proposal for the development of an integrated international legal instrument for the application of the PIC procedure, the phasing out of POPs and any additional international measures to promote chemical safety. Since the Copenhagen meeting was not a negotiating session for the PIC instrument, there was no formal EU position which the Commission was mandated to present. Those EU member states that participated in the meeting acted in their own capacity. While no consensus could be reached on a formal recommendation to the Governing Council to consider the integrated approach as proposed by the Netherlands and Belgium, due primarily to opposition from the United States and Australia, the report of the meeting invited the Executive Director of UNEP to request the views of governments on such an approach and submit them to the nineteenth session of the Governing Council for consideration (UNEP 1996c: 9).

The controversy over the scope of the PIC instrument was not resolved during the second session of the INC/PIC in Nairobi. The EU reiterated and further formalised its position by proposing that the objective and scope of the instrument be formulated in general terms and that provision be made for the possible adoption of protocols containing additional measures to achieve this objective. The United States and like-minded OECD countries continued to oppose any extension of the objective and scope of the instrument beyond PIC. The G77, which had not taken any clear position on this issue during the first meeting, remained hesitant to enter the fray and did not directly address the EU proposal. It did, however, indicate that the PIC procedure *per se* was insufficient and that the instrument should therefore include a reference to the more general objective of promoting the environmentally sound management of chemicals, especially in developing countries, in accordance with the principle of 'common but differentiated responsibility' (UNEP 1996b). Thus formulated, the objective of the instrument would justify the inclusion of detailed provisions on capacity-building and technical assistance, which seems to be the primary goal of the G77. However, the concept of 'environmentally sound management of chemicals' is sufficiently broad to be able to encompass also other measures additional to the PIC procedure, as envisaged by the EU. While there seems to be room for compromise between the EU and G77 positions, it remains to be seen whether common ground can be found with all OECD countries.

Leaving aside the fundamental political controversy with respect to the scope of the PIC convention, the work of the INC/PIC so far in translating the existing voluntary instruments into legally binding treaty provisions has shown that, even within the narrow confines of these instruments, a number of contentious points remain to be solved. One controversial question is that of export notification. All developing countries that spoke on the issue, as well as the EU, strongly supported the inclusion of export notification provisions in the PIC convention, while the United States, Australia, Canada and Japan argued that such provisions, though included in the existing voluntary UNEP and FAO instruments, were 'not appropriate for discussion in this forum' (UNEP 1996a: 7, para. 29) because they are not part of the PIC procedure *stricto sensu*. Export notification mechanisms were also criticised for being potentially trade-restrictive.

The US position on this issue is somewhat puzzling, since export notification requirements for pesticides not registered for use in the United States and for certain other toxic chemicals subject to regulatory restrictions have been in force under US domestic law since the late 1970s.[19] Apparently, the United States does not want to assume any international obligations in this field unless it can be assured that other exporting parties will be subject to equally onerous notification requirements. The US feels that export notification, as currently practised, places a disproportionate burden on US exporters, since the EU requires export notification only for those chemicals banned or severely restricted *under EU law*, while a large number of hazardous chemicals subject to bans or restrictions in individual EU member states are not covered by any export notification requirements.

The EU, for its part, has shown a similar concern about ensuring a 'level playing field' in another area of the negotiations. While, under the EU's 1992 regulation, as pointed out above, exporters are legally required to comply with PIC decisions of importing countries participating in the voluntary UNEP/FAO PIC procedure (ECM 1992: art. 5, para. 4), there is no similar legislation in other OECD member states. Thus, the EU has not only a political, but also a direct trade interest in making sure that the PIC convention requires all exporting parties to impose the same compliance obligation on exporters under their jurisdiction. Countries which view the PIC procedure as essentially an information exchange mechanism, and consider that the enforcement of PIC decisions is the sole responsibility of importing countries, are reluctant to commit themselves to any form of export control. This is apparently why the United States, at the first meeting of the INC, opposed the inclusion in the PIC convention of a provision requiring exporting parties to take appropriate enforcement measures. At the end of the second meeting, however, the EU's view seemed to prevail, as a consensus appeared to be emerging on draft language committing each exporting party to 'take appropriate legislative and/or administrative measures to ensure compliance by exporters of prior informed consent chemicals in its territory' (UNEP 1996b: 19, art. 10c).

A range of other, more technical, issues remains to be solved, particularly as regards the criteria and procedure for the inclusion of chemicals in the PIC procedure. As the voluntary procedure is being transformed into a legally binding one, there is a need to further elaborate and formalise both the substantive and procedural provisions governing the listing of PIC chemicals. In this process, there is a strong tendency, encouraged by some chemical-exporting countries, to raise the threshold for the inclusion of chemicals by increasing the number of procedural steps and the amount of data to be provided, introducing additional substantive and formal criteria and maximising the room for discretionary review of national regulatory decisions which may qualify a chemical for PIC listing (UNEP 1996b: Annex I, art. 6–8, Annex Y).

At the nineteenth session of the UNEP Governing Council that took place in Nairobi from 27 January to 6 February 1997, there was considerable debate on the relationship between the PIC negotiating process and other chemical safety issues currently on the international agenda. Acting on behalf of the EU, the Dutch presidency tabled a draft decision on chemicals management in which it proposed to initiate 'an expeditious assessment process on the advantages and disadvantages of a framework convention on chemicals based on chapter 19 of Agenda 21'. The EU draft decision also called for an early conclusion of the negotiations on the PIC instrument and for the establishment of an intergovernmental negotiating committee to develop an international legally binding instrument on POPs, and urged that both the PIC and POPs conventions should have 'the necessary flexibility to allow for easy adaptation in view of future developments relating to the different instruments on the safe and environmentally sound management of chemicals' (UNEP 1997a). After protracted negotiations, the Governing Council reached a compromise on a package of four distinct but related decisions on chemicals management issues.

The EU failed in its bid to have the Governing Council direct the INC/PIC to incorporate 'the necessary flexibility' in the PIC instrument. Instead, the Governing Council 'confirm[ed] the present mandate' of the INC, while at the same time 'recogniz[ing] that *additional elements* relating to the prior informed consent procedure are under consideration' in the Committee (UNEP 1997b), thus avoiding any direct reference to the notion of 'further measures' as referred to in its earlier decision 18/12, and in the report of the Copenhagen expert meeting. However, the Governing Council adopted a separate decision on 'further measures to reduce the risks from a limited number of hazardous chemicals', in which it endorsed most of the recommendations of the Copenhagen meeting, while merely *noting* those concerning 'possible bans or phase-outs', but nevertheless inviting the IFCS to 'consider taking action, as appropriate, to implement them and to report on such action' to the 1999 session of the Governing Council (UNEP 1997c).

The same session will also consider a report to be drawn up by UNEP's Executive Director, 'outlining options of both a legal and administrative nature for enhanced coherence and efficiency among international activities related to chemicals' (UNEP 1997d) including the PIC convention, which is to be adopted and opened for signature at a diplomatic conference to be convened in early 1998, and the POPs instrument, which is to be developed by a new INC which will start its work in the same year (UNEP 1997e). One of the 'options' to be examined in the Executive Director's report, though not explicitly mentioned, will obviously be the EU's proposal for a framework convention, but the Governing Council will only be able to review these options at a time when the PIC convention will already have been signed and when the negotiations on a separate POPs convention will be well under way.

Conclusions

For many years, EU policy on export of chemicals was guided by purely economic considerations, with minimal regard for the environmental implications within importing states. EU legislation applied only to the internal market and contained no restrictions on the export of hazardous chemicals. The rise of environmental concern within the EU – particularly the unilateral actions taken by a few of the 'greener' member states – created pressure to reform EU legislation in a manner which reconciled environmental concerns with sensitivity for the EU's competitiveness in global chemical export markets. Although the Commission proposed in the mid-1980s that the EC should take the lead in regulating international trade in banned and severely restricted chemicals, the Council balanced environmental with economic concerns by seeking international harmonisation rather than pushing ahead with tough unilateral export restrictions.

Internal regulation has resulted in a united EU external negotiating position versus other OECD states who are seeking to maintain their preferred policy approaches, which confer competitive advantages in export markets. Having unilaterally made the prior informed consent procedure legally binding for EU exporters, the EU now has an obvious interest in the adoption of a multilateral legal instrument making this procedure mandatory for all exporting states. In the current international negotiations, the EU is pressing for a convention which will not only achieve that objective, but also provide a flexible legal framework for further international cooperative action to control hazardous chemicals, beyond the PIC procedure. So far, the EU has demonstrated a remarkable level of internal cohesion and effectively articulated its common position in these negotiations. It is unlikely, however, that the EU will be able to achieve all its negotiating objectives in the face of strong opposition from other OECD countries and G77 scepticism. The outcome of the negotiations will be indicative of the success of the EU's ambitious effort to lead the development of international law in the field of chemical safety.

Notes

1 In 1978, EC countries accounted for 61.5 per cent of world pesticide exports (Bull 1982: 6). Since export statistics are not available on a substance-by-substance basis, it is impossible to distinguish exports of banned and severely restricted pesticides from total exports.

2 In 1985, seven international NGO networks formed the *Coalition Against Dangerous Exports* (CADE) to lobby for controls on hazardous exports from the EC. See Chetley (1985).

3 See, e.g. Council Directive of 26 June 1978 (78/631/EEC) on the approximation of the laws of the member states relating to the classification, packaging and labelling of dangerous preparations (pesticides), OJ, 1978, L 206/13, art. 1, para. 2 (c); Council Directive of 21 December 1978 (79/117/EEC) prohibiting the placing on the market and use of plant protection products containing certain active substances, OJ, 1979, L 33/36, art. 5 (b). The rationale for this exemption, as spelled out in the preamble of Directive 79/117/EEC, is that 'it is not appropriate to apply Community provisions to plant protection products intended for export to third countries *since in general these countries have their own regulations*'.

4 Based on figures published by the Central Statistical Office of the Netherlands in 1986 (CBS 1986: 5).

5 For a discussion of national legislative measures in the Netherlands, Germany and the United Kingdom, see Pallemaerts (1987). For an overview of these intergovernmental initiatives, see Pallemaerts (1988).

6 Ironically, the preamble of an OECD Council recommendation on exports of hazardous wastes adopted a few weeks before the Commission issued its proposal explicitly refuted this parallelism by stressing 'the particular nature of wastes and *the distinction between wastes and products which are traded internationally*' (OECD 1986, emphasis added).

7 Similar objectives were espoused by the Commission in the field of marine pollution. See Chapter 4.

8 Article 113 confers broad powers on the Community in the field of the 'common commercial policy' which covers all aspects of external trade with non-member countries, including, *inter alia*, 'export policy'. As early as 1979, the Court of Justice of the European Communities, in an important opinion on the scope of the common commercial policy, held that Article 113 could serve as a basis for 'a commercial policy aiming at a regulation of the world market for certain products rather than at a mere liberalisation of trade'. See ECJ (1979).

9 The legal basis of legislation plays a crucial role in several of the cases under investigation in this volume. Unanimous voting has prevented the adoption of an EU carbon–energy tax (see Chapter 5), whereas the possibility of qualified majority voting facilitated the adoption of EU measures restricting methyl bromide (see Chapter 2).

10 A first draft proposal for amendments was presented by the Commission to national experts at the first meeting of the designated authorities for the application of Council Regulation No. 1734/88 on 22–23 June 1989 (EC 1989).

11 Three chemicals were added to the list at the time of adoption of the new regulation in 1992, and fifteen more in December 1994. (See ECM 1994.)

12 It should be noted that the Council went beyond the Commission's initial proposal in making the labelling and packaging requirements applicable to *all* exported dangerous chemicals. The Commission had originally proposed that they should apply only to the chemicals listed in Annexes I and II of the regulation, i.e. those subject to the export notification and/or PIC procedure. (See EC 1991: art. 6, para. 1.)

13 The first entries in Annex II were very incomplete at the time of their adoption, as they covered only six PIC chemicals and a limited number of participating importing country decisions. Annex II was updated and completed only two and a half years later (EC 1996).

14 See the discussion below regarding the UNEP/FAO Intergovernmental Negotiations.

15 On the establishment and early activities of the Working Group, see generally Sankey (1989).

16 Although the official GATT documents do not identify the single contracting party that blocked the adoption of the draft decision, it is a matter of public knowledge that it is the United States. See O'Connor (1992: 8), Ward (1994: 278, fn. 60).

17 Decision 15/30, see Pallemaerts (1990–1994).

18 See also proposals submitted by the Group of 77 and China (UNEP 1995b).

19 Toxic Substances Control Act (TSCA), s. 12(b), 15 USC. § 2611(b); Federal Insecticide, Fungicide and Rodenticide Act (FIFRA), s. 17(a), 7 USC § 1360(a).

Bibliography

AE (1987) *Agence Europe*, No. 4673, 4 December.

BISD (1983) GATT Basic Instruments and Selected Documents, 29th Supplement, March.

—— (1990) GATT Basic Instruments and Selected Documents, 36th Supplement, Decision of 19 July 1989.

Bull, D. (1982) *A Growing Problem: Pesticides and the Third World Poor* (Oxford: Oxfam).

CBS (1986) Centraal Bureau voor Statistiek, *Kwartaalbericht Milieu*, 1986/2.

Chetley, A. (1985) *Cleared for Export: An Examination of the European Community's Pharmaceutical and Chemical Trade* (Brussels: CADE).

EC (1986a) Doc. COM(86) 362 final (Communication from the Commission to the Council).

—— (1986b) Proposal for a Council Regulation Concerning Export from and Import into the Community of Certain Dangerous Chemicals, 25 June 1986, OJ 1986, C 177/5.

—— (1989) EC Doc. XI/155/89-Rev. 1.

—— (1991) Proposal for a Council Regulation (EEC) Concerning the Export and Import of Certain Dangerous Chemicals, Doc. COM(90) 591 final, OJ 1991, C17/16.

—— (1993a) Commission of the European Communities, *Informing the Importer – Guide to Council Regulation EEC/2455/92 Concerning the Export and Import of Certain Dangerous Chemicals*, Luxembourg, May.

—— (1993b) Report of the Commission on the Implementation of Regulation (EEC) No. 1734/88, Doc. COM(93) 121 final, 31 March.

—— (1994) Commission Regulation (EC) No. 41/94 of 11 January 1994 amending Annex II of Council Regulation (EEC) No. 2455/92 Concerning the Export and Import of Certain Dangerous Chemicals, OJ 1994, L 8/1.

—— (1996) Commission Regulation (EC) No. 1492/96 of 26 July 1996 amending Annexes II and III of Council Regulation (EEC) No. 2455/92 Concerning the Export and Import of Certain Dangerous Chemicals, OJ 1996, L 189/19.

ECJ (1979) Opinion 1/78, 4 October 1979, *International Agreement on Natural Rubber*, [1979] ECR 2909.

ECM (1986) Council Directive of 12 June 1986 (86/279/EEC) amending Council Directive 84/631/EEC on the Supervision and Control within the European Community of the Transfrontier Shipment of Hazardous Waste, OJ 1986, L 181/13.

—— (1988a) Council Regulation (EEC) No. 1734/88 of 16 June 1988 Concerning Export from and Import into the Community of Certain Dangerous Chemicals, OJ 1988, L 155/2.

—— (1988b) Council Resolution of 16 June 1988, OJ 1988, C 170/1.

—— (1992) Council Regulation (EEC) No. 2455/92 Concerning the Export and Import of Certain Dangerous Chemicals, OJ 1992, L251/13.

—— (1994) Council Regulation (EC) No. 3135/94 of 15 December 1994 amending Annex I of Council Regulation (EEC) No. 2455/92 Concerning the Export and Import of Certain Dangerous Chemicals, OJ 1994, L 332/1.

EP (1981a) Reply of the Commission to Written Question No. 1370/80 of Mr Glinne, OJ 1981, C 49/15.

—— (1981b) Reply of the Commission to Written Question No. 2162/80 of Mr Michel, OJ 1981, C 140/15.

—— (1982a) Reply of the Commission to Written Question No. 648/82 of Mrs Poirier, OJ 1982, C 245/10.

—— (1982b) Reply of the Commission to Written Question No. 1082/82 of Mr Rogers, OJ 1982, C 339/10.

—— (1983a) Report drawn up on behalf of the Committee on Environment, Public Health and Consumer Protection on the Export of Various Dangerous Substances and Preparations (Rapporteur: Mrs V. Squarcialupi), EP Doc. 1–458/83 (27 June 1983).

—— (1983b) Deb. European Parliament, OJ 1983, Ann. No. 1–304 (Statement of Commissioner Narjes).

—— (1983c) EP Res. on the Export of Various Dangerous Substances and Preparations, 14 November 1983, OJ 1983, C 307/109.

GATT (1988a) GATT Doc. MTN.TNC/8(MIN), December.

—— (1988b) GATT Doc. MTN.TNC/W/14.

—— (1988c) GATT Doc. MTN.GNG/W/18, 17 November.

—— (1989) GATT Doc. L/6467, 2 February.

—— (1990) GATT Doc. DPG/W/9, 17 April.

—— (1991) Draft Decision on Products Banned or Severely Restricted in the Domestic Market, GATT Doc. L/6872, 2 July.

—— (1994) Decision on Trade and Environment, adopted by Ministers at the Meeting of the Trade Negotiations Committee in Marrakesh, 14 April 1994, in *The Results of the Uruguay Round of Multilateral Trade Negotiations* (Geneva: GATT), pp. 469–471.

MoP (1994) Decision II/12 of the Second Meeting of the Conference of the Parties to the Basel Convention on the Control of Transboundary Movements of Hazardous Wastes and their Disposal, Geneva, 21–25 March.

ND (1983) Memorandum from the Netherlands delegation, EC Council Doc. 7289/83 (Annex), 30 May.

O'Connor, B. (1992) 'GATT and the Environment,' *Review of European Community and International Environmental Law*, Vol. 1, pp. 6–13.

OECD (1985) Conclusions and Recommendations of the OECD Conference on International Co-operation concerning Transfrontier Movements of Hazardous Wastes, Basel, Switzerland, 26–27 March.

—— (1986) OECD Council, Decision-Recommendation on Exports of Hazardous Wastes from the OECD Area, C(86)64(final), 5 June.

Pallemaerts, M. (1987) 'Export Notification: The EC Approach in the International Context,' *European Environment Review*, Vol. 1, No. 2, pp. 25–37.

—— (1988) 'Developments in International Pesticide Regulation,' *Environmental Policy and Law*, Vol. 18, pp. 62–68.

—— (1990–1994) 'Regulation of Chemicals: The Year in Review,' *Yearbook of International Environmental Law*, Vol. 1 (1990), pp. 163–166, Vol. 2 (1991), pp. 170–175, Vol. 3 (1992), pp. 281–287, Vol. 4 (1993), pp. 205–212, Vol. 5 (1994), pp. 209–218.

Sankey, J. (1989) 'Domestically Prohibited Goods and Hazardous Substances – A New GATT Working Group is Established,' *Journal of World Trade*, Vol. 23, No. 6, pp. 99–108.

TSCN (1983) *Toxic Substances Control Newsletter*, Autumn.

UN (1994) Report of the Commission on Sustainable Development on the Work of its Second Session, UN Doc. E/CN.17/1994/20, 12 July.

UNEP (1995a) Governing Council of the UNEP 18th Session, Programme Committee, 'Proposed Amendments to the Draft Decision on the Development of an Internationally Binding Instrument for the Application of the Prior Informed Consent Procedure for Certain Chemicals in International Trade, Proposals submitted by France and Malaysia,' UN Doc. UNEP/GC.18/PC/CRP.22, 18 May.

—— (1995b) UN Doc. UNEP/GC.18/PC/CRP.24, 19 May.

—— (1995c) UNEP Governing Council decision 18/12, 26 May.

—— (1996a) 'Report of the Intergovernmental Negotiating Committee for an International Legally Binding Instrument for the Application of the Prior Informed Consent Procedure for Certain Hazardous Chemicals and Pesticides in International Trade on the Work of its First Session,' UN Doc. UNEP/FAO/PIC/INC.1/10, 21 March.

—— (1996b) 'Report of the Intergovernmental Negotiating Committee for an International Legally Binding Instrument for the Application of the Prior Informed Consent Procedure for Certain Hazardous Chemicals and Pesticides in International Trade on the Work of its Second Session,' UN Doc. UNEP/FAO/PIC/INC.2/7, 12 November.

—— (1996c) 'Report of the Government-designated Group of Experts on Further Measures to Reduce the Risks from a Limited Number of Chemicals,' UN Doc. UNEP/GC.19/INF.7, 21 November.

—— (1997a) Governing Council of the United Nations Environment Programme, 19th Session, Committee of the Whole, 'Chemicals Management, Netherlands: Draft Decision Submitted on Behalf of the States Members of the European Union,' UN Doc. UNEP/GC.19/CW/CRP.3, 28 January.

—— (1997b) UNEP Governing Council Decision 19/13A, 7 February.

—— (1997c) UNEP Governing Council Decision 19/13B, 7 February.

—— (1997d) UNEP Governing Council Decision 19/13D, 7 February.

—— (1997e) UNEP Governing Council Decision 19/13C, 7 February.

Ward, H. (1994) 'Trade and Environment in the Round – and After,' *Journal of Environmental Law*, Vol. 6, pp. 263–295.

Weir, D. and M. Schapiro (1981) *Circle of Poison: Pesticides and People in a Hungry World* (San Francisco: Institute for Food and Development Policy).

WTO (1996a) WTO Doc. WT/CTE/W/32, 30 May.

—— (1996b) Report of the Committee on Trade and the Environment, 7 November (Geneva: WTO).

4

IMPROVING COMPLIANCE WITH THE INTERNATIONAL LAW OF MARINE ENVIRONMENTAL PROTECTION

The role of the European Union

André Nollkaemper

In recent years, the European Community (now Union) has expressed grand ambitions to improve compliance with the international law of marine environmental protection.[1] This ambition responds to two pressures. Poor compliance with international standards has contributed to degradation of the European marine environment and deterioration of the competitiveness of the Community's maritime sector. While in many respects the objectives of promoting environmental protection and economic competitiveness are competing, the attempt to ensure world-wide adherence to gradually rising international standards may serve both objectives at the same time.

The European Community (EC) has not been well positioned to improve compliance with international laws, however. In contrast to its role vis-à-vis other international environmental issues, like depletion of the ozone layer (Jachtenfuchs 1990) and transboundary movement of hazardous wastes (Allen 1995), the role of the Community vis-à-vis the international law of marine environmental protection has been a modest one. Member states have chosen to develop and implement international law without Community involvement. They have restricted the international role of the EC to a bare minimum. The question is whether in the face of continuing pressure on the marine environment and competitiveness, member states will depart from this practice and allow the EC to develop a more significant role.[2]

In this chapter I will discuss the development of the EC policy that aims to improve compliance with the international law of marine environmental protection. I will examine the pressures for reform as perceived by the

Community, the actual reforms taken and proposed by the EC, and the limitations on such reforms imposed by European and international law. The chapter concludes that the role of the EC in reforming international law remains a modest one and is likely to stay so; positions of member states and third states as well as the legal system are not conducive to an enhanced role. In the second section I will describe the background of the Community policy to improve compliance with international law. The third section sets out the narrow legal margins within which the Community has to seek an independent role to contribute to improved compliance of international law. In the fourth and fifth sections I will examine the main policies carried out or proposed by the EC to serve simultaneously the objectives of environmental protection and economic competitiveness. The final section will summarise the leading threads.

The pressures for reform

The compliance-gap

The body of globally applicable laws that aims to protect the worlds' oceans from pollution is massive and diverse. It consists of dozens of treaties and an incalculable number of non-treaty instruments, including many resolutions adopted by the International Maritime Organization (IMO).[3] It covers such wide-ranging issues as routing schemes, standards for crews, technological standards for separation of waste water, allocation of jurisdictional entitlements of coastal and flag states, and liability rules.[4]

However, by now it is clear that these international laws have not fully achieved their objectives. While international laws leave little to be desired in terms of substance, and formally are adhered to by most shipping nations, their implementation has been less than perfect (Peet 1992, Nollkaemper 1993: 555, Donaldson 1994). Sub-standard shipping is pervasive: the shipboard and shore-based management, levels of crew-training and maintenance of many ships are not in accordance with internationally agreed rules (OECD 1996: 5).

Ever more ships are registered in states that do not adequately enforce international rules, many of them being so-called 'open register' states (that is, states which do not limit eligibility for registration to ships whose owners or operators have any connection with that state). Lord Donaldson's authoritative Inquiry into the prevention of pollution from merchant shipping notes that the vice of open registers is twofold. First, they lead to sub-standard levels of safety. Second, they have led some ship owners to shop around the world for registers which have the lowest standard of enforcement and which, in consequence, involve them in the least expense (Donaldson 1994: 61). Flag states with particularly high losses include Malta, Turkey, Cyprus, South Korea, Vanuatu, Panama, Greece and the

Philippines. All these states except the last have expanding fleets (Donaldson 1994: 62). The variation in loss levels is high: the worst being 100 times the best in fleets of 2 million or more (EC 1993a: 7). Also the annual reports of the port state control authorities provide information as to laggards – the top ten deficiency ratios per flag state include Romania, Honduras, Syria, Turkey, Cuba, Belize, Egypt, Morocco, Algeria and Lebanon (MoU 1995: 47). Within the EC figures are slightly better, but variations between the best and the worst are still high; in terms of loss rate the worst (Greece) is still fifty times the best, whereas deficiency ratios vary between 52 and 12 per cent (EC 1993a: 7–9). Recent figures for the United Kingdom confirm these data. In 1995, the Marine Safety Agency of the United Kingdom claimed it had taken 'the world lead in publishing details of ships detained'. Over 10 per cent of 1,821 foreign ships inspected in UK ports in 1995 failed to comply with international standards. States with the highest percentages of detained ships per inspected ships were Turkey, Honduras, Bulgaria, India, Malta and Panama. EC 'toppers' were Greece and Denmark.[5]

The causes of non-compliance are manifold. But one factor appears to be the key driving force behind inadequate compliance: non-compliance pays. This is most evident for the ship owner, who has the ultimate responsibility for the safe and pollution-free operation of a ship. Many ship owners seek the margin below international standards and above the level required for operation of the ship in order to improve their competitiveness. In recent years, ship owners have sought this margin by cutting investments in the maintenance of ships, understaffing crews and reflagging their ships to registers in states that impose less costly standards. Such decisions are facilitated by other actors. Many flag states[6] accept or even stimulate sub-standard performance. There is a consistent pattern of sub-standard ships operating under the flags of countries that have ratified relevant treaties. The OECD has found that inadequacy of flag state surveys is 'largely due to the fact that ship registration remains for these flag states, primarily, a competitive business and that commercial considerations are often seen to override safety matters' (OECD 1996: 18; see also Donaldson 1994, Chapter 6). Also certain classification societies, hired to guarantee compliance with standards on board ships, play a role. To improve their market share they take only 'soft' glances at safety standards (OECD 1996: 20).[7] Insurers and financiers also play their part, as 'many flag states, class societies, insurers, financial institutions do not feel any responsibility for the safety performance of the shipping industry and optimize private profits disregarding social costs, aided by other players' (Nieuwpoort and Meijnders 1996). The interests of none of these actors are one-dimensional, though. As many flag states are also port states and coastal states, their possible commercial interests in promoting or supporting sub-standard shipping may be neutralised by their 'environmental interests' in keeping their coasts free from pollution, or by a more general environmental concern, irrespective of a state's own coast.

The economics of the sector are unfavourable to improving the above situation. Over-capacity is vast, freight rates have tumbled, and profit margins are low or negative (Nieuwpoort and Meijnders 1996: 9). The spiral is downward. More ships have chosen to register in open registers to achieve benefits such as tax allowances, freedom to crew ships with low-wage labour, regardless of nationality and without involvement of labour unions, less stringent vessel classification and inspection rules, and to search for standards of operation that make the ship sail, yet do not incur the high costs required for compliance with international standards. The resulting pattern of inadequate compliance imposes two pressures on the European Community and its member states. First, sea transport continues to pose threats to European marine ecosystems. Second, the equality of competition that European states sought to achieve by adhering to international standards is achieved only on paper.

Environmental and competitiveness pressures

Poor implementation of existing treaties contributes to the continuing threats the maritime sector poses to marine ecosystems. Deliberate intentional discharges from ships as well as shipping accidents occur regularly in European waters. While shipping is responsible for only 12 per cent of all marine pollution, the effects remain a major cause of concern in the waters of the member states of the EC and the world over, not so much in terms of their global impact but certainly in terms of their impact on local coastal marine ecosystems. Affected interests include those of seabirds, fish, marine mammals, local communities dependent on fisheries, and public health (Donaldson 1994: 25–26). Though some of the casualties in European waters are caused by ships registered in member states of the EC, it appears that responsibility for the majority of these cases lies with ships registered outside the EC. Illustrative is that Lord Donaldson's Inquiry reports the incidents which led or could have led to significant pollution during the course of its work – twenty-seven of those cases involved ships registered outside the EC, whereas twelve involved ships registered in the EC (Donaldson 1994: Appendix L). The 1995 Annual Report of the Memorandum of Understanding on Port State Control is even clearer. It lists the top twenty-two flag states, ranked in percentage of ships delayed or detained following inspection. This list includes only one member state (Portugal) (MoU 1995: 47). The top twenty-two of priority cases for inspection in 1996/1997 (based on detention percentages) include only two member states: Greece and Portugal (MoU 1995: 53). As one can assume that the controls on the basis of the Memorandum of Understanding expose which ships pose a threat to European waters, this clearly indicates that the main threats to European waters come from ships flying the flag of non-member states. It is thus clear that the EC suffers environmentally from poor compliance in EC states, but in particular in non-EC states.

Failure to observe international requirements for safety and environmental protection not only causes environmental concerns. It also means that ship owners and flag states that fail to comply with international standards enjoy competitive advantages.

In most European states, the number of ships flying the national flag has decreased radically. In 1970, 32 per cent of the world tonnage remained under the flag of EC member states. In 1994, this figure had decreased to 14 per cent (EC 1996: 6). While for certain countries, including Luxembourg and Greece, ship owners have flagged in, the trend in the EC member states is that ship owners are flagging out. For instance, the number of ships sailing the Dutch flag has decreased from 548 (1986) to 371 (1993), while the number of Dutch vessels sailing a foreign flag increased in the same period from 251 to 365 (Asteris 1992, Peeters *et al.* 1996). The EC and the member states have perceived the demise in their shipping industry as a reason for concern. A shipping sector is still thought to be important for strategic and security reasons. In addition the economic losses are considerable in terms of loss of employment, loss of fiscal revenues, and loss of work in ship-related industries (EC 1996: 6).

The root cause of the trends in European shipping are due to competitiveness differences. For the main part, these appear to be caused by high labour costs and fiscal regimes (Asteris 1992: 192–194, Aspinwall 1996: 138–140). In addition, however, there is some evidence that part of the competitiveness loss is due to the fact that European ships operate in unfair international competition, as flags of convenience states are unable or unwilling to enforce international standards for safety and environmental protection.[8] The very reason why states have relied on globally, rather than regionally, applicable treaties to regulate environmental effects of shipping is to ensure that environmental regulations do not undermine the competitive position of individual states and ship owners. Failure to comply with agreed rules undermines this objective and may threaten competitiveness. Poorly maintained and staffed ships compete with unfair advantage with those which have made the additional financial efforts to meet international requirements.

Generalisations on the financial benefits of ship owners and operators who choose not to comply with international standards are difficult. Everything depends on the age of the vessel, the type of trade it is engaged in and the abilities of the crew. However, a recent OECD report provides evidence that clearly indicates that 'financial advantages linked to the non-observance of agreed international rules and standards can reach a substantial amount of the running costs of a vessel' and that by 'deliberately [avoiding] compliance with international rules and standards which govern safety and pollution prevention . . . ship owners obtain financial benefits which can equate to a significant percentage of total vessel operating costs' (OECD 1996: 5, 9).

The OECD report substantiates that there exists a considerable financial difference between the minimum level of expenditure to ensure owners' compliance with international standards, on the one hand, and the so-called 'floor level', which corresponds to the minimum expenditure to keep the vessel operational, on the other. With an assumed standard level of operation for a twenty-year-old bulk carrier, for instance, standing at $3,250 a day, the margin of sub-standard operation could be as big as $500 a day or $182,500 a year. This, so the OECD report indicates, represents 13 per cent of the annual running costs for this type of vessel (OECD 1996: 10–15).

Cost savings can be enjoyed in a variety of ways, ranging from decisions not to put on board the required oil-spill clean up requirements or the required fire-fighting equipment, putting officers on board who do not hold valid licences, or using defective navigational equipment. All such decisions directly or indirectly contribute to increasing the risk of adverse environmental effects of shipping operations. While, again, generalisations are difficult, the above figures indicate the economic benefits of non-compliance and the potential impact on competitiveness. These benefits will grow with the development of ever more safety and environmental requirements.

The Community's response to the compliance-gap

The above developments thus confronted the Community with two policy targets: improving environmental protection and improving competitiveness. In recent decades, member states had already initiated a wide variety of policy responses. These do not necessarily further both objectives at the same time. Unilateral Community standard setting for pollution control may further environmental protection, and possibly competitiveness within the Community, but undermine competitiveness world-wide. Haralambides (1993) argued that tougher environmental standards in the shipping sector will, in the short run, 'undoubtedly interfere with competitiveness'. However, he adds that tougher standards also tend to induce innovation that may lessen costs in the long run.[9] As I will further indicate below, this is not the policy presently pursued by the Community. Only on relatively marginal issues has the Community proposed or implemented unilateral measures (that is, apart from the IMO standard setting process) to improve safety and environmental protection, and the possibility of a win-win situation on this point has not been explored.

Conversely, competitiveness may be improved by reducing safety standards to the minimum level required by international law, thus curtailing environmental performance. As noted above, considerable costs may be saved in the margin between 'good practice' and 'standard practice' (OECD 1996: 10). This is presently being considered by the Netherlands. The Netherlands imposes certain requirements that go beyond the requirements of the applicable international laws (Peeters *et al.* 1996: 112–113). A recent economic analysis conducted for the government estimated that the costs for ship owners of requiring environmental performance that exceeds international requirements may amount to 1.3–1.6 per cent of annual operation costs (ibid.). The Netherlands Government announced that they will consider closely which requirements may be lowered so as to save costs. Although not all of those costs concern requirements directly related to environmental safety, they do indicate the type of policy responses searched for by states anxious to protect and improve competitiveness.

Alternatively, government intervention could subsidise ship owners or ports, thereby offsetting the competitive effects of stringent unilateral environmental protection and thus insulating the industry from the effects of competitive disadvantages.[10] Such measures might include subsidies, fiscal incentives and imposing cargo reservations for national flag carriers. Europe has attempted to avoid government-imposed cargo reservation on trade between developed countries (Asteris 1992: 195). On the other hand, member states, sanctioned by the Commission, have actively supported shipping industries by providing state aid. While this, *prima facie*, is prohibited by the EC Treaty, the Commission has considered that 'the importance of maintaining and developing the shipping sector for economic and employment reasons as well as the particular nature of the international competition it faces' justify exceptions (EC 1996: 28). Another often used measure to promote competitiveness is the creation of specific registers for ships flying their flag to alleviate competitive disadvantages, in particular vis-à-vis open registers outside the EC. These registers were created to exclude ships flying the flag of the member state from certain costs in the fiscal and labour regime of the normal ('first') register. This has been done in Germany, Denmark and Finland (Aspinwall 1996: 136, EC 1996: 10).

While most of these policy options are presently being considered, their success has been limited. The Commission of the EC noted that they 'have not been able to reverse the flagging out' (EC 1996: 11). It is against this background that the Community has with full force entered a new direction: to ensure that safety and environmental standards are applied equally throughout the world, thus undermining much of the competitive advantages flag states and ship owners seek in the margin below international standards and above the minimum level needed to keep a ship in operation. It is this policy that unites the twin objectives of improved competitiveness and environmental protection.

The choice to pursue improved compliance as a key objective signifies a radical departure from existing policies. For a long time, the Community did not adopt an active policy to enhance compliance with international laws for marine environmental protection. The EC confined itself to support for the development and entry into force of rules in the IMO and their implementation in the member states, e.g. by recommending that member states ratify the Solas[11] and Stcw[12] Conventions (Erdmenger 1991).[13] It did not incorporate international laws into Community law and did not bother to attempt to enforce international laws through its own Community law procedures.

This low intensity regulation prevailed until the early 1990s. The accident with the *Braer* off the Shetland Islands in 1992 formed a watershed. The Community realised once more that implementation of international shipping law was utterly inadequate. The Community now deemed that it could no longer rely exclusively on the efforts of member states and the IMO to improve compliance. On 8 June 1993, the Council adopted the legislative plan of work on safety and environmental aspects of shipping laid down in the 'Common Policy on Safe Seas'.[14]

The key objective of the Common Policy on Safe Seas is to contribute to the protection of European waters. However, in the shade of this objective the Common Policy on Safe Seas also seeks the enhancement of competitiveness. The policy repeatedly makes the point that uneven implementation of treaties has undermined the competitive equality that was the rationale of the global system of laws (EC 1993a: para. 22). The Common Policy on Safe Seas aims to enhance safety and prevention of pollution at sea through the elimination of sub-standard operators, vessels and crews from Community waters, and thereby seeks the combined objectives of protection of the European waters and the protection of competitiveness (EC 1993a: 2). This policy is confirmed in the later document 'Towards a New Maritime Strategy' – the policy document that aims to restore the economic position of the European shipping industry. This states that 'it is the conviction of the Commission that the strict enforcement of a safety policy based on internationally agreed standards will lead to a marked improvement of the competitive situation of ships under EC registers with stringent safety enforcement' (EC 1996: 12).

The choice for improving compliance appears less sensitive and controversial than most of the other policy choices to improve environmental performance and/or alleviate competitive disadvantages, such as subsidies, cargo reservations, or unilateral substantive standards that go beyond international standards and hope to achieve a win-win situation. Nonetheless, the role of the EC to further the objective of improved compliance is by no means unproblematic, as I will discuss further below.

I will examine the main elements of the Community policy to reduce the compliance-gap of international law for marine environmental protection. First, however, I will discuss the legal margins that determine the leeway for the Community to contribute to implementation of this body of international law.

The narrow legal margins for Community action

The leeway for the Community to expand its role vis-à-vis the international law for marine environmental protection is strongly determined by applicable rules of European and international law. I will discuss these under three headings: the principle of subsidiarity, the Community's external competence in this particular field, and the applicable rules of international law.

Subsidiarity

The provisions on both common transport policy and environmental policy provide the Community with ample authority to develop a policy for the implementation of international laws applying to ship-based marine pollution.[15] However, the exercise of this competence is subject to the principle of subsidiarity as laid down in Article 3B of the Maastricht Treaty. Under this principle, the Community shall only take action if and insofar as the objectives of the proposed action cannot be sufficiently achieved by the member states.

As noted above, for many years member states did not use the Community institutions and implicitly considered that Community action was not necessary. However, in 1993 the Commission considered that prevailing reliance on policies of member states, either individually or through international institutions, was no longer adequate. The Common Policy on Safe Seas argues that individual action by member states has not produced adequate results in the past and is unlikely to do so in the future. We can assume that this means that the Commission finds reliance on implementation by individual member states, as well as on international policies of member states through the IMO, ineffective. The Community, so the Common Policy on Safe Seas continues, thanks to its political and legislative machinery, would be uniquely placed both to ensure that member states apply standards to ships flying their flags in a more uniform and rigorous manner, and to enforce respect for the same standards on vessels of all flags when operating in EC waters (EC 1993a: para. 31).

The subsidiarity principle does not provide a clear-cut answer to whether Community intervention is indeed necessary. As a practical matter it is difficult or even impossible to examine whether more or less Community action is necessary to achieve the objective of environmentally safe sea transport. It is clear that international law has *not* fully yielded the envisaged results. Whether EC intervention will prove more effective is uncertain at best. The subsidiarity principle is a two-edged sword which cannot only cut against Community action but also against state action (Fischer 1994, Golub 1996). The adoption of the Common Policy on Safe Seas clearly indicates a slight shift in the perceptions of member states in favour of Community intervention. However, as I will discuss further below, the scope of this shift and the depth of Community intervention remain a controversial matter. Whether or not Community action is necessary will have to be decided on each issue by an application of the broadly formulated subsidiarity test, that thus only sets fairly general parameters for the development of Community policy vis-à-vis the international law of marine environmental protection.

External policy

The EC policy on ship-based marine pollution is largely an external policy that aims not only to regulate ships flying the flags of member states but also the flags of non-member states. As in all external environmental policies, the competence of the Community remains a hotly debated issue (Leenen 1984, Temple Lang 1986, Nollkaemper 1987, Haigh 1991, Brinkhorst 1994, Hession and Macrory 1994).

There is no doubt that in principle the EC is competent to pursue an external policy regarding marine environmental protection law.[16] These competencies in principle are concurrent:[17] while the Community is competent to accept and negotiate commitments under relevant treaties, whether the Community will do so is up to the member states. But when the EC has *sole* authority to act, and member states have to refrain from acting, is a question that continues to bewilder practitioners and scholars. The Council has never laid down a clear allocation of external powers in general or with regard to marine environmental protection, and has deferred to the European Court of Justice (ECJ) to fill the legal vacuum. The current position is that on most issues, the Community and member states share concurrent powers, a situation that calls for application of the ambiguous doctrine of cooperation (ECJ 1993: para. 5). The Common Policy on Safe Seas refers in many places to the need to coordinate positions in conformity with this doctrine.

The unanswered question is: what are the legal and environmental consequences of the obligation to cooperate? What does this obligation mean in light of the prevailing practice in international institutions where coordination meetings often result in cacophony?[18] Opinions can be found on any position between two extremes – either it is said that the duty to cooperate implies that member states must take up a common position, and that if no common position can be agreed upon, member states should refrain from acting unilaterally (Frid 1995: 149); or it is said that the duty to cooperate does not amount to an obligation to reach a common position, but only to an obligation to use best endeavours to do so (MacLeod *et al.* 1996, Neuwahl 1996). It appears that for all practical purposes, the effect of the duty of cooperation will depend on a pragmatic, case-by-case seeking for coordinated positions.[19] Like the principle of subsidiarity, the law on the external powers of the Community only provides broad parameters within which member states and Community institutions have to bargain for acceptable allocation of roles.

There is no uniform EC policy in international institutions related to marine environmental protection vis-à-vis non-member states. Member states more often than not pursue policies outside the EC context. In recent years coalitions of different groupings of member states of the EC have been more common than positions taken by the EC as a whole against non-member states. Often these coalitions are separated by a north–south line (with northern states generally being more supportive of stringent environmental policies than southern states with big fleet interests, notably Greece), but other coalitions also occur, much depending on the issue. The point is that, in contrast to fisheries, for instance, the law is not sufficiently developed to support and mandate a coherent external EC policy. Of course, this does not imply that a change in the law (more exclusive powers for the EC) will *in itself* change these coalitions – such a change in the law will only develop once the member states' perceptions of their interests converge, thus allowing more Community legislation and corresponding external powers.

International law

A third set of determinants of Community action is derived from international law. International law imposes certain limits on the competence of the Community to enforce standards for ships sailing the flags of non-member states when these ships are outside the exclusive economic zone, territorial sea or ports of member states (Nollkaemper and Hey 1995). In several respects international law limits the possibilities for a further development of policies of port states and coastal states vis-à-vis ships sailing the flag of non-member states. For instance, it has often been suggested that IMO routing schemes could be more effective if they were mandatory. However, coastal states cannot do so unilaterally, because EC or national law cannot exceed international standards. Outside their territorial sea states cannot impose rules and enforce them against ships sailing the flag of a third state.[20] A Community rule that would make such resolutions mandatory could not be enforced against ships sailing the flags of non-member states. Ships sailing the flags of third states can only be bound with consent of the flag state, which in practice will mean after IMO approval. Other Community policies that are hampered by international legal considerations are the proposed directive concerning the setting up of a European vessel reporting system (that would require ships entering or passing through maritime zones to report to port authorities) in the maritime zones of the member states[21] and the Proposal for a Council Directive on the Minimum Level of Training for Maritime Occupations (COM(93)217).

However, the limits imposed by international law are easily overstated. While they impose certain limits on Community action, this does not mean that they paralyse the Community. They leave room for unilateral action. For instance, international law permits the Community to oblige its member states to apply safety and environmental standards that are more stringent than international law to ships flying their flag; permits the Community to oblige member states to improve their exercise of port state jurisdiction, by which it can indirectly affect ships flying the flags of non-member states; and permits the Community to apply its relatively effective enforcement procedures to supplement the renowned inefficacious enforcement procedures of international law. It is up to the Community to seek and explore the margins for unilateralism and, through that, contribute to the change of international law.

Improving implementation of international law

Convergent implementation of international law by member states

A first line of the EC policy to improve implementation of international law is targeted at the environmental performance of ships flying the flags of its member states. The Community considered that one reason for the poor implementation of international laws was the 'soft nature' of such laws – either in terms of their contents or in terms of their legal nature (EC 1993a: para 35).[22] The use of ambiguous terms and the use of recommendations would not sufficiently guarantee environmental protection and equality of conditions of competition. The Community attempts to take away this assumed source of poor implementation by developing internal Community rules that fill gaps in the applicable international laws.

This policy suggests that the Community finds that environmental performance of the ships of member states is sub-optimal, and that the European waters and coasts are threatened by pollution from European ships. As noted earlier, the reports of the Memorandum of Understanding on Port State Control provide only limited support for this assumption. Although performance of certain flag states (notably Portugal and Greece) is troublesome, the threats posed by these ships to European waters is negligible compared to those posed by non-EC ships. Nonetheless, the results of port state control do suggest that at least some ships flying flags of member states do not comply with their international obligations. By approving of the Common Policy on Safe Seas, the EC member states accepted the reasoning of the Commission that environmental performance of member states should be improved.

The EC policy on convergent implementation also suggests, though this point is not elaborated in the EC's policy documents, that intra-Community competitiveness is jeopardised by different interpretation and application of international law (EC 1993a: 17). To the knowledge of this author, there are no studies of such effects. Although there is of course competition between neighbouring ports (like Rotterdam and Hamburg), most observers consider that this is more a matter of overall quality of the port, the connections between the port and subsequent transport facilities, and possibly dues for entry into port (an issue not regulated by international law), than a matter of differences in the application of international law or even unilateral substantive requirements.

Nonetheless, the EU has made a priority the development of internal Community rules that are to harmonise standards and enforcement in its member states. This represents a break with existing policy. Until the early 1990s, the Community had adopted few internal laws, dealing mainly with the control and reduction of pollution caused by hydrocarbons discharged at sea.[23] Otherwise it totally relied on support for rules developed in the IMO. However, the Community has not totally abandoned its reliance on the work of the IMO. Most of its (proposed) internal rules are not fully unilaterally imposed environmental and safety standards that go beyond what is agreed in IMO, but rather measures that aim to strengthen IMO rules with a view to furthering their implementation. The Community has taken a threefold approach.

First, the Community attempts to achieve convergent application of generally formulated obligations contained in IMO Conventions. It attempts to substitute detail for vagueness. An example is the development of safety performance standards for marine equipment.[24] Ship-borne equipment is covered by treaties, mainly SOLAS and MARPOL. The Common Policy on Safe Seas notes that despite these international rules, different levels of national standards exist, implementing the international rules or recommendations concerning technical specifications and testing procedures, leaving margins of appreciation to national bodies. This would lead to different levels of safety *despite the existence of international standards*. One result is that 'shipping companies face higher costs in some member states than in others because of differing national requirements, and accordingly are at a competitive disadvantage to companies in other member states' (EC 1993a: Part II, para. 16). For these reasons the Commission has proposed a directive that harmonises technical requirements and testing procedures.

Second, the EC aspires to insert legally binding rules in areas that are now governed by resolutions. The IMO has used many Resolutions in the implementation and specification of treaties. Apparently, there is a large number of variations in approaches of member states for the implementation and interpretation of legally non-binding standards. The EC has deemed it appropriate to remedy this situation by pursuing a mandatory application of IMO Resolutions by transforming them into directives.

An example was IMO Resolution A747(18) on segregated ballast tanks (SBTs). SBTs have certain environmental advantages. In conventional oil tankers the same tanks are used for water when in ballast and for cargo when laden. When ballast water is discharged, oily waste enters the marine environment. Tankers equipped with SBTs separate the ballast water from the cargo and in principle ballast waters do not contain oily wastes. However, many ships have been reluctant to use SBTs as ports discriminate against

SBT tankers: because they are bigger and ports tend to calculate dues according to overall tonnage, SBTs are charged more than less environmentally friendly tankers (EC 1993a: paras 7.54–7.62). While some ports (including Rotterdam) had unilaterally reduced port dues for SBTs, this is not the global pattern and presently there is an economic disincentive for using SBTs. In order to avoid owners who order SBT tankers paying additional port costs, the IMO has called for a non-binding resolution on states to follow the example of Rotterdam. Reportedly, however, this resolution has been poorly implemented. The Commission noted that ports 'in many countries, including some Member States' continue to charge SBT tankers in a manner which does not comply with this resolution (EC 1993a: 32). The EC now aims to make the IMO resolution mandatory within the EC and hopes to contribute to solving this problem (EC 1993a: para. 7.57). In the future more IMO resolutions may be transformed into directives, including, for example, the International Maritime Dangerous Goods Code (IMDG) and the Code for Construction and Equipment of Ships Carrying Dangerous Chemicals in Bulk (EC 1993a: Part II, paras 27–28). Although no comprehensive data are available that support these figures, as noted above, the results of port state control do suggest that at least some ships flying flags of member states (including Greece, Portugal, Spain and Denmark) do not comply with their international obligations. By approving the Common Policy on Safe Seas, the EC member states accepted the reasoning of the Commission that implementation by member states of the more important IMO Resolutions should be improved.

Third, the Community has initiated additional rule-making to ensure that member states fulfil the obligations flowing from internationally agreed standards on safety and environment. For instance, it has proposed common criteria for registers that would have to ensure environmental protection and be conducive to eliminating distortions of competition which can result from varying registration conditions and flag state enforcement (EC 1996: 15–16). Also, the Community has adopted a directive on classification societies (Directive 94/57) that aims to improve the quality of the work of these societies so as to ensure that ships supervised by member states as flag states live up to international standards.

The EC policy to enhance the convergent implementation of international law by member states may induce better compliance. One obvious result is that the EC institutions can improve compliance by invoking the enforcement procedures available under the EC Treaty. Although a case can be made that the Community could enforce treaties directly, that is, without transformation into Community law (Nollkaemper and Hey 1995), no use has been made of this possibility. In any case, the possibilities of enforcing legally non-binding IMO rules will be limited. Transforming international rules into Community law may increase the use of enforcement proceedings and thereby compliance in member states.

However, in several respects the policy of the Community aimed at the convergent implementation of international law is limited. First, whether the mere fact that a resolution is adopted in legally binding form and in a more detailed manner actually will enhance compliance is a matter of some uncertainty. It is clear that international law has *not* fully yielded the envisaged results. But whether that is due to the fact that IMO rules are vague and often non-legally binding is uncertain at best. Whether legally binding and detailed EC law will prove more effective is equally uncertain.

Second, this part of Community policy will have a narrow scope – it will mainly affect ships flying the flags of member states. From an environmental perspective the gains may be limited, as it affects only ships of member states, while, as noted earlier, most threats are caused by ships from non-member states. From the perspective of competitiveness, such unilateral action undermines the very uniformity that was the *raison d'être* of the global rules. Even though these EU rules are not fully unilateral, as they are based on international rules, rather than prescribing new rules, they are inspired by the idea of improving environmental performance within the Community, thus enlarging rather than reducing the compliance-gap. For this reason it is necessary that unilateral Community action targeted at ships flying the flags of member states is accompanied by policies targeted at ships flying the flags of non-member states.

Strengthening enforcement by port states

A second line of the EC policy to improve implementation of international law is to improve port state enforcement. The role of port states is critical in enforcement of international laws. Ideally, they perform only a marginal role as the prime responsibility for enforcement should rest with flag states. However, inadequate performance of flag states has put the task of port state to the fore. Port states can step in, as a form as self-defence, and try to remove sub-standard ships from their ports. This may serve the indirect aim of pressing flag states and ships to improve environmental performance and thereby to restore this aspect of competitiveness.

The Law of the Sea Convention sanctions port state control, whereas the IMO Conventions rely heavily on port state control to ensure that vessels conform to the safety, technical, environmental and social standards for seafarers as laid down in the conventions. In the early 1980s, European states developed the concept of port state control in the Paris Memorandum on Understanding on Port State Control (MoU 1982). Signatories agreed to harmonise and increase inspection of compliance with seven treaties. Practice since then has shown that the problems of sub-standard shipping are real: the number of deficiencies found by authorities has remained at a constant high level. Port state inspection is now widely regarded as 'a very, if not the most, important policing mechanism for the shipping industry', which acts as a safety net for flag state compliance control (OECD 1996: 23).[25]

Port state control can simultaneously contribute to the objectives of environmental protection and competitiveness. Port state authorities inspect deficiencies, and can call on a ship to remedy such deficiencies; prevent a vessel from proceeding on its voyage; and notify the flag state and next port of such deficiencies. Such measures can make it more difficult for ship owners to obtain financial benefits by not observing international rules. Depending on the tightness of control measures and corrective actions, port state control can minimise or offset the financial advantages resulting from the operation of sub-standard vessels.[26]

The role of the EC in port state control over compliance with international law has remained relatively marginal. In the early 1980s, the Council rejected a proposal for a directive on the harmonisation of port state control[27] – member states were concerned that this could trigger an external competence of the EC that could detract from their competencies in the IMO (Erdmenger 1991: 1191–1192),[28] and opted for the development of the Memorandum on Port State Control as an instrument outside the Community framework. Since then the Community has played a modest role as a member of the MoU.

However, the Common Policy on Safe Seas envisaged a new role for the Community. The high number of sub-standard ships that continue to operate in Community waters; the number of detections, which varies greatly between countries participating in MoU; and the lack of uniformity in inspection criteria and detention criteria would, so the Common Policy on Safe Seas argues, call for Community intervention. By developing its own policy on port state control, the EC hopes to improve compliance with international law of ships flying the flags of non-member states.

In 1994 the Community adopted the directive of port state control.[29] By adopting this instrument the Community has offered a modest contribution to port state control of compliance with international law. In a number of respects this Directive differs from the MoU:[30] it displays more detail, including minimum requirements for inspection; it is legally binding, and consequently enforceable in the Community. There are no signs that the Community intends to impose additional substantive requirements on ships entering ports, as has been done by the United States. Rather than imposing new requirements, the Community attempts to use international law as a weapon against non-member states. Its policy fits in with the prevailing pattern where port states are reluctant to use their economic power to impose new regulations because of fear of retaliations or loss of trade to rival ports (Nieuwpoort and Meijnders 1996: 11).

It must be noted that all this is largely based on unexpressed and unsupported assumptions that a different policy (that is, one that would impose substantive requirements) would indeed lead to loss of trade to rival ports.

Although in the northern European context there is considerable competition between ports, use of substantive requirements for entry into ports has never been a key policy to compete with other ports in the region. Likewise, there is no hard evidence that the introduction of port state control in Europe (the same would hold for the EC) has caused the operation of substandard ships to shift to other regions. However, as one expert added, the assumption that this may be the case as such 'seems valid enough' (Schifferli 1993: 442). It is for this reason that European states have for some time, and with increasing success, promoted the development of port state controls in other parts of the world.

Strengthening international law

A third envisaged Community contribution to the improved compliance of international law is the effective participation of the Community in competent international organisations, particularly the IMO. Long before the EC recognised the ineffectiveness of existing laws to ensure environmental protection and competitiveness, contracting parties of the relevant IMO Conventions had done so. During the past decade(s), they have initiated a continuous process of reform. The present agenda includes many items critical to improvement of compliance, most of all improvement of flag state performance (including development of guidelines and better assistance to developing countries); extension of port state control; and clarification and improvement of vessel traffic services. While, as noted above, the EC is pursuing a unilateral policy on some of these issues, there is no doubt that if the EC really wishes to achieve progress in environmental protection and competitiveness, it also has to reach ships flying the flags of non-member states. It is for this reason that a key strategy of the Common Policy on Safe Seas is to improve contents and implementation of international law by supporting developments in the IMO.

Effecting this objective requires a drastic change in prevailing practice of external Community policy. For a long time, the Community's role in IMO has been limited. It is an observer, but not a party to the IMO Conventions. This is no oversight: member states have not wished the Community to perform that role. As previously mentioned, by 1980 the existence of external powers of the Community was already a matter of discussion. A proposal for a directive on the harmonisation of port state control was rejected by the Council. Illustrative also is the sorry plight of proposals for accession to the London Dumping Convention (Suman 1991).

The Commission has envisaged a surge in the use of external powers. In the Communication on the Common Policy on Safe Seas as well as in other documents it refers to the exercise of external powers as an important means of enhancing safety and environmental protection. It has called on member states to 'boost co-operation' and envisages that the EC 'participate as contracting party in all new conventions (maritime safety, prevention of pollution) so as to secure consistency' (EC 1993b: 18). Most of all this is relevant to efforts to improve environmental performance of flag states. Thus, the Commission proposes a joint effort by Community and member states in the IMO to agree on a world-wide basis on certain conditions for flag administrations and their ship registers (EC 1996: 14–15); and has called on member states to provide support to the Sub-Committee on Flag-State Implementation; to provide input in regulatory discussions and support to developing countries (EC 1993a: Action Programme annex para. 126); and effectively to coordinate action to help strengthen IMO action on the role of the human element (EC 1993a: Action Programme annex paras. 128–137).

The prospects of this policy remain uncertain. As noted above, European Community law only defines in broad terms the scope of competence of the Community. Member states and Community institutions will have to elaborate on a case-by-case basis whether they will coordinate their positions and proceed through the Community institutions. The Community cannot, under present Community law, force its member states to surrender their role in the IMO to the Community. The key question then is whether member states are willing to give up their traditional freedom in IMO. In the 1980s member states were not willing to do so, and resisted almost any exercise of external powers by the Community. Recent years have shown substantial development in the willingness of member states to act in a co-ordinated manner within a Community framework. Coordination between the member states and the Commission for regulatory discussions in the IMO has become standard procedure. However, the Community retains a low profile. As noted above, the coalitions between member states in the IMO are manifold, and member states, in differing coalitions, oppose each other more often than they act together in the EC vis-à-vis non-member states. Observers consider that this can be explained by a variety of factors: general environmental awareness/priorities (which explains the common coalition of northern European states), position towards desirability of government control (which puts the United Kingdom often in the coalition of southern states), and the more narrow factor of flag states' interests combined with low environmental concern (of which Greece is the best example). But, again, much depends on the issue at stake, and most observers feel generalisations are not helpful. As yet, it remains unlikely that the Community would actually replace member states.

Concluding observations

Inadequate compliance with international law for the protection of the marine environment imposes continuing pressure on the marine environment as well as on the competitiveness of the European shipping sector. The

EC has thus tried to achieve the joint objectives of improving environmental protection and safeguarding competitiveness by improving international compliance. To this extent, global treaties remain critical to the objectives of world-wide environmental protection and competitiveness.

The Community tries to pursue these joint objectives by three strategies: convergent implementation in member states; port state control; and external action within the framework of the IMO and other relevant international institutions. On each of these points, the Community attempts to use the margins left open by European and international law. As yet, however, these three policies have led to limited results – the Community has adopted a few directives that intend to undo the ambiguity of international law: a directive on port state control that largely overlaps with the Paris MoU; and a cautious attempt to improve use of Community institutions in the framework of the IMO. Member states do not appear eager to grant the Community an independent role vis-à-vis the international law of marine environmental protection. The Community's policy has been defensive rather than aggressive – it aims to defend and improve compliance with existing laws, but is reluctant to develop new standards unilaterally. Whereas it may wish to promote new standards internationally, its legal and political status in the IMO makes the prospect for doing so limited.

Notes

1 I use the term 'law of marine environmental protection' to refer to the body of international law applying to the environmental effects of maritime transport. I will not refer to the laws applying to marine pollution from other sources.

2 The Commission's role in representing the Community, and the tension this creates with the ability of member states to negotiate individually, is also discussed in Chapters 2, 3 and 5.

3 Established under the Convention on the Intergovernmental Maritime Consultative Organization, Geneva, 6 March 1948, 9 UST 621.

4 The main treaties concluded within the framework of the IMO relevant to the present chapter include the International Convention on the Safety of Life at Sea, 1 November, 1974, (1975) 14 *International Legal Materials* 959 [hereinafter the SOLAS Convention]; the International Convention on Standards of Training, Certification and Watchkeeping for Seafarers, 5 July, 1978, (1984) *UKTS* 50, Cmnd 9266 [hereinafter the STCW Convention]; and the International Convention for the Prevention of Pollution by Ships, 2 November, 1973, (1973) 12 *International Legal Materials* 1319 [hereinafter the MARPOL Convention]. The United Nations Convention on the Law of the Sea, 10 December, 1982, (1982) 21 *International Legal Materials* 1261 [hereinafter the Law of the Sea Convention] includes the core jurisdictional rules relevant to regulation of ship-based marine pollution. For an overview of relevant international laws, see Donaldson (1994, Chapter 7).

5 Department of Transport, 'New figures show 190 ships held in UK ports', 29 June 1995, available on Internet: http://www.coi.gov.uk/depts/GDT/coi8358a.ok.

6 Under art. 94 of the Law of the Sea Convention, a flag state is obliged to exercise its jurisdiction and control over ships sailing its flag.

7 At present a two-tier market in classification society has been created, with societies in the second tier performing poorly. See also Donaldson (1994), para. 7.30.

8 The Government of the Netherlands expressly identified this as an important cause of unfair competition; see Letter of the Minister of Transport, Water Management and Public Works to Parliament of 23 May, 1995, K. 1994–95, 24165, nr 1, p. 5. See also EC (1996), Chapter B.II.

9 Such win-win strategies are discussed in the introductory chapter of this volume, and play an important role in EU policy to reduce greenhouse gas emissions (Chapter 5).

10 All government interference leads to market distortions, as it subsidises price dumping and over-capacity, and puts more pressure on freight rates and thus fuels the need for more cost-reduction by ship owners (Nieuwpoort and Meijnders 1996). The use of such 'carrots' is discussed in the introductory chapter and has played a particularly important role in relation to climate change (Chapter 5), greening the CAP (Chapter 7) and EU development aid (Chapter 8).

11 Recommendation 78/584, OJ 1978 L 194/17.

12 Recommendation 79/114, OJ 1979 L 33/33.

13 Accidents in 1979 involving the *Aragon* and the *Khark-V* induced the Council to adopt a Resolution that calls on member states to ensure stricter compliance by ships flying their flags with the technical rules for safety and pollution prevention contained in, in particular, the Marpol and Solas Conventions; and to intensify the inspections of foreign ships in their ports, as provided for in the MoU on Port State Control, particularly focusing on compliance with the manning requirements of the Solas and Stcw Conventions. OJ 1990 C 206/1.

14 Council Resolution on a common policy on safe seas, OJ 1993, C 271/1.

15 The EC Treaty provides sufficient legal basis for the development of safety and environmental rules on shipping. Such rules are primarily to be based on art. 84(2). It may be argued that art. 75 (in the case of safety rules) and art. 130s (in the case of environmental rules) can be used as additional legal bases; but in principle art. 84(2) is a sufficient legal basis and can incorporate the objectives of art. 75 and 130s. See the discussion of legal bases in Chapter 3.

16 Art. 84(2), the only provision expressly providing for regulation of maritime transport, does not make any mention of external powers. However, there is little doubt that the Commission possesses implied external powers under the case-law laid down in ECJ (1971, para. 16) (hereinafter ERTA), ECJ (1976, paras 19–20), ECJ (1993, paras 7–9). Article 130r(4) does provide for an external competence in regard to environmental policy.

17 Concurrent powers are powers that the Community may exercise if the Council so decides, but which are not (yet) exclusive Community powers; see Temple Lang (1986).

18 On the inadequacy of existing coordination in many international institutions, see Maunu (1995: 126).

19 This might change, though, if the institutions were to impose more precise obligations of cooperation on member states (Neuwahl 1996: 678).

20 Within their territorial seas coastal states can prescribe and enforce routing systems as long as these are compatible with the right of innocent passage. But there is some room for discussion as to what specific information states may require before infringing on the right of innocent passage. See Plant (1990).

21 Proposal for a Council Directive concerning the setting up of a European vessel reporting system in the maritime zones of Community Member States, OJ 1994 C 22; amended proposal OJ 1994, C 193. The draft directive contains a number of provisions that should safeguard the rights of non-member states. Art. 2(3) provides that the application of Eurorep shall be 'without prejudice to the right of inoffensive passage through territorial waters, the right of unhindered passage in transit through international straits, and freedom of navigation outside territorial waters'. Art. 5(3) provides that the reporting obligation of art. 5(1) is not mandatory for transit vessels (that is, vessels not bound for Community ports or intending to anchor in territorial waters or flying the flag of a member state). After the entry into force of the Solas amendments, mandatory reporting for vessels carrying dangerous or polluting goods will be permitted.

22 For a discussion of the various ways in which international obligations can be 'soft', see Reisman (1988).

23 Resolution of 26 June 1978 setting up an action programme on control and reduction of pollution caused by hydrocarbons discharged at sea, OJ 1978 C 162. This action programme responded to the infamous stranding of the *Amoco Cadiz* off the coast of Brittany in 1978. The Council also set up an Advisory Committee and an information system relating to the combating of oil pollution. See Commission Decision 80/686 (OJ 1980 L 188), amended by Decisions 85/208 (OJ 1985 L 89) and 87/144 (OJ 1987 L 57), and Council Decision 86/85 (OJ 1986 L 77/33).

24 Proposal for a Council Directive on marine equipment, OJ 1995, C 218/9.

25 However, port state control has been criticised as an 'end-of-pipe' solution that attempts to solve economic problems by fighting only symptoms and does not create incentives for the necessary self-organisation in the industry. See Nieuwpoort and Meijnders (1996: 10).

26 Costs of corrective action should be high enough to compensate the daily financial benefits. The OECD calculates that in cases involving both detention and non-detention the daily benefits may still outweigh the costs of corrective action (OECD 1996: 16–17, 23).

27 OJ 1980 C 192/8.

28 Subsequently, the draft directive provided the basis for the MoU on port state control.

29 Directive 95/21 concerning the enforcement, in respect of shipping using Community ports and sailing in the waters under the jurisdiction of the member state, of international standards for ship safety, pollution prevention and shipboard living and working conditions, OJ 1995, L 157/1.

30 See discussion of the MoU and the role of the EU by Molenaar (1996).

Bibliography

Allen, M. (1995) 'Slowing European Hazardous Waste Trade: Implementing the Basel Convention into European Union Law,' *Colorado Journal of International Environmental Law and Policy*, Vol. 6, pp. 163–182.

Aspinwall, M. (1996) 'The Unholy Social Trinity: Modelling Social Dumping Under Conditions of Capital Mobility and Free Trade,' *West European Politics*, Vol. 19, pp. 125–150.

Asteris, M. (1992) 'British Shipping in Decline,' in M. Asteris and P. Green (eds), *Contemporary Transport Trends* (Aldershot: Avebury), pp. 189–197.

Brinkhorst, L. J. (1994) 'The European Community at UNCED: Lessons to be Drawn for the Future,' in Deirdre Curtin and Ton Heukels (eds), *Institutional Dynamics of European Integration. Essays in Honour of Henry G. Schermers, Vol. II* (Dordrecht: Martinus Nijhoff).

Donaldson, Lord (1994) *Safer Ships, Cleaner Seas. Report of Lord Donaldson's Inquiry into the Prevention of Pollution from Merchant Shipping* (London: HMSO).

EC (1993a) Commission of the EC, A Common Policy on Safe Seas, COM(93)66 final, 24 February.

—— (1993b), Commission of the EC, 'Community Participation in International Organs and Conferences,' SEC(93)36, 1 March.

—— (1996), Commission Communication 'Towards a New Maritime Strategy,' COM 96(81) final, 14 March.

ECJ (1971) Case 22/70, *Commission v. Council*, [1971] European Court Reports 263.

—— (1976) Case 3, 4, and 6/79, *Kramer et al.*, [1976] European Court Reports 1279.

—— (1993) Opinion 2/91, *ILO Convention 170*, OJ 1993, C 109/1.

Erdmenger, Jürgen (1991) 'Artikel 84,' in H. von der Groeben *et al.* (eds) *Kommentar zum EWG-Vertrag Volume 1*, fourth edn (Baden-Baden: Nomos).

Fischer, T. (1994) '"Federalism" in the European Community and the United States: A Rose by Any Other Name,' *Fordham International Law Journal*, Vol. 17, No. 2, pp. 389–440.

Frid, R. (1995) *The Relations Between EC and International Organisations. Legal Theory and Practice* (The Hague: Kluwer Law International).

Golub, J. (1996) 'Sovereignty and Subsidiarity in EU Environmental Policy,' *Political Studies*, Vol. 44, No. 4, pp. 686–703.

Haigh, N. (1991) 'The European Community and International Environmental Policy,' *International Environmental Affairs*, Vol. 3, No. 3, pp. 163–180.

Haralambides, H. (1993) 'A New Future for European Shipping,' inaugural address, available from Erasmus University, Rotterdam.

Hession, M. and Macrory, R. (1994) 'The Legal Framework of European Community Participation in International Environmental Agreements,' *New Europe Law Review*, Vol. 2, pp. 59–136.

Jachtenfuchs, M. (1990) 'The European Community and the Protection of the Ozone Layer,' *Journal of Common Market Studies*, Vol. 28, No. 3, pp. 261–277.

Leenen, A. Th. S. (1984) 'Participation of the EEC in International Environmental Agreements,' *Legal Issues of European Integration*, pp. 93–111.

MacLeod, H., I. Hendry and S. Hyett (1996) *The External Relations of the European Communities* (Oxford: Clarendon Press).

Maunu, A. (1995) 'The Implied External Competence of the European Community after the ECJ Opinion 1/94: Towards Coherence or Diversity?', *Legal Issues of European Integration*, pp. 115–128.

Molenaar, E. (1996) 'The EC Directive on Port State Control in Context,' *International Journal of Marine and Coastal Law*, Vol. 11, pp. 241–288.

MoU (1982) Memorandum of Understanding on Port State Control, Paris, 1982, reprinted in David Freestone and Ton Ijlstra (eds) (1990) *The North Sea: Perspectives on Regional Environmental Co-operation* (London: Graham and Trotman).

—— (1995) Memorandum of Understanding on Port State Control, Annual Report.

Neuwahl, N. (1996) 'Shared Powers or Combined Incompetence? More on Mixity,' *Common Market Law Review*, Vol. 33, pp. 667–678.

Nieuwpoort, G. and Meijnders, E. L. M. (1996) 'An Integration of Economic and Safety Policy for Shipping. The Need for Self-organisation,' unpublished paper on file with author.

Nollkaemper, A. (1987) 'The European Community and International Environmental Co-operation – Legal Aspects of External Community Powers,' *Legal Issues of European Integration*, pp. 55–91.

—— (1993) 'Agenda 21 and Prevention of Sea-based Marine Pollution: a Spurious Relationship?', *Marine Policy*, Vol. 17, pp. 537–556.

Nollkaemper, A. and E. Hey (1995) 'Implementation of the LOS Convention at Regional Level: European Community Competence in Regulating Safety and Environmental Aspects of Shipping,' *International Journal of Marine and Coastal Law*, Vol. 10, pp. 281–300.

OECD (1996) *Competitive Advantages Obtained by Some Shipowners as a Result of Non-observance of Applicable International Rules and Standards* (Paris: OECD).

Peet, G. (1992) 'The MARPOL Convention: Implementation and Effectiveness,' *International Journal of Estuarine and Coastal Law*, Vol. 7, pp. 277–295.

Peeters, C. *et al.* (1996) *De Toekomst van de Nederlandse Zeevaartsector* (Delft: Delft University Press).

Plant, G. (1990) 'International Legal Aspects of Vessel Traffic Services,' *Marine Policy*, Vol. 14, pp. 71–74.

Reisman, W. M. (1988) 'A Hard Look at Soft Law,' *Proceedings of the American Society of International Law*, Vol. 82, pp. 373–337.

Schifferli, R. (1993) 'The Memorandum of Understanding on Port State Control: Its History, Operation, and Development,' in A. Cooper and E. Gold (eds), *The Marine Environment and Sustainable Development: Law, Policy, and Science* (Honolulu: Law of the Sea Institute), pp. 434–447.

Suman, D. (1991) 'Regulation of Ocean Dumping by the European Economic Community,' *Ecology Law Quarterly*, Vol. 18, pp. 559–618.

Temple Lang, J. T. (1986) 'The Ozone Layer Convention: A New Solution to the Question of Community Participation in "Mixed" International Agreements,' *Common Market Law Review*, Vol. 23, pp. 157–176.

5

THE PATH TO EU CLIMATE CHANGE POLICY

Thomas Heller

Although debate about the physical science of global warming continues, it is increasingly clear that the centre of gravity of the climate change controversy is moving from science to politics (Bruce *et al.* 1996, Houghton *et al.* 1996, Watson *et al.* 1996). Nowhere is this shift more visible than in the European Union (EU), even though neither Europe's contribution to the problem of climate change, nor its risks therefrom, would seem to demand a leading engagement. In 1987, the twelve nations of the European Community emitted only 12 per cent of the CO_2 emissions from fossil fuels. The United States was responsible for 22 per cent; the bloc of the centrally planned economies 36 per cent (EC 1992a: 9). By 1993, the EU-12 accounted for 14.5 per cent of the global total. The accession of Austria, Sweden and Finland raised the EU share to over 16 per cent. But the United States still produces more than 25 per cent and the now depressed economies of Central Europe and the former Soviet Union over 17 per cent.

The greatest part of the growth in emissions arises in the developing world whose proportional share, if current trends are unabated, will rise from its present 25 per cent to more than 50 per cent by 2020–2030 (Bergesen *et al.* 1994: 9–10). The levels of per capita CO_2 emissions in the EU are in the lower range of other OECD nations, similar to those of Japan and well below those of the United States, Canada, Australia and New Zealand (ibid.: 11). Finally, the likely prospects of direct damages from climate change are not particularly dire in the European Union (Bruce *et al.* 1996: 205). In some Scandinavian nations, there may be positive effects of warming on leisure and agriculture. European oil producers, like the United Kingdom and Norway, will suffer sectoral export losses (Sydnes 1996: 270–272). In other EU member states, such as Holland, the costs of adapting known technologies to deal with the incremental effects of rising sea levels may be relatively low compared to losses portended in the Third World.

In spite of this seeming lack of exceptional scientific or economic salience of climate change in Europe, the political role of the European Union and its member states has been central to the efforts to shape an international regime for climate change. Between 1987 and 1992, in the face of ascendant green domestic interests and in the moral afterglow of the successful multilateral negotiation of the Montreal Protocol on Substances that Deplete the Ozone Layer (ILM 1987), European nations made the initial commitments to stabilise or reduce emissions of greenhouse gases (Beuermann and Jäger 1996, Sydnes 1996: 193–195, 269–270, 276–278). On 13 June 1990, Germany, the largest emitter in the EC-12, agreed to a 25 per cent cut by the year 2005 from its 1987 baseline in the former West Germany. This decision was reaffirmed by the government in November 1990 when projections of still greater reductions in the new Bundesländer were added. On 29 October 1990, the joint Council of EC Energy and Environment Ministers agreed to stabilise CO_2 emissions by the year 2000 at 1990 levels (Bergesen *et al.* 1994: 17). In 1992, the European Union became a party, as a regional entity, to the United Nations Framework Convention on Climate Change (FCCC) and alone among OECD nations aggressively sought to have the developed economies assume in the treaty stabilisation objectives like those of the 1990 EC resolution. Yet, the years since the Rio accord have been deceptive for those with faith in Europe's forward stance.

Writing about the contested German implementation of its reduction targets in words that could be applied to the Union as a whole, analysts have noted 'a contradiction between Germany's claim to be a leader in climate change politics at the EU level and in the CoP (Conference of the Parties to the FCCC) process and the development of national climate policy' (Sydnes 1996: 223). Bold European programmes to mitigate climate change with new policy instruments, centred on the enactment of a Community level carbon–energy tax, stalled in disagreements over their economic consequences for employment and international competitiveness. Those emissions reductions that have been recorded in the member states are less the outcomes of legislative initiatives aimed at climate *per se*, than the by-products of recession, German reunification and industrial restructuring. The political force of climate change has been secondary to that of economic recovery, and realism about prospective costs has clouded the environmental idealism of Europe's original pronouncements. The defeat of the carbon–energy tax has underlined the growing recognition that Europe's ability to fulfil its climate change pledges will turn on the discovery of a strategy that lowers the total costs of mitigation to a tolerable political level. In this context, it is logical that there should be renewed attention to proposals for emissions trading which assume there is a wide cost variance in mitigation opportunities among different nations and that trading will make least cost options available generally to participants in a unified market.

There are a variety of forms that emissions trading may take. The most familiar are the cap and trade systems employed by the Clean Air Act of 1990 for sulphur dioxide trading in the United States and the little used, but similarly structured, CFC trading provisions in the Montreal Protocol (Heller 1996: 297, 325). In each of these cases a global limit on emissions is negotiated and a quantity of permits equal to this cap is allocated by some formula or auction to emitting organisations. Permit holders may then trade them freely so that the market will ensure that the cheapest emissions reductions options will always be those that are exercised (Tietenberg 1985, Tripp and Dudek 1989, Dudek and Wiley 1994: 334–350). Joint implementation (JI) is a somewhat ill defined variant on a pure cap and trade market that has come to the fore of climate change politics because of its mention in FCCC Article 4.2.[1] In some discussions, the concept of JI seems to refer to any system of emissions trading. In others, JI refers more specifically to an offset system for foreign investments that lead to emissions reductions and is viewed as a precursor to a full trading market in fungible emissions permits. In this more limited sense, a typical JI transaction would allow actors with obligations to limit emissions in their home jurisdiction to net against their legal obligations the emissions savings they finance with their investments abroad.

There are numerous controversies over the structure and effects of a JI system that arise in the multilateral context where only some parties to an international climate change regime have been assigned emissions caps (Heller 1995: 17–30, 1996: 301). What needs concern us here is that either JI or more comprehensive emissions trading markets can reduce the costs of mitigation obligations to the extent that they widen the pool of low cost reduction options. As such, it holds out the possibility that Europe can escape its current climate change impasse. The remainder of this chapter explores the logic of these constraints and possibilities. The first part recalls the sequence of the proclamation of European leadership in climate change, the design of a European programme to implement this claim, and the political failure to effect it. The second part suggests that Europe's interest in trading systems will only be successful in getting beyond this failure if it envisions the market to include extra-European jurisdictions. This implies that Europe will have to revise the international regime building strategy it has so far championed in the multilateral (FCCC) forum to rescue emissions trading from the political deadlock in which those negotiations are now stalled.

The European road towards trading: 1985–1996

Defining the leading role

Europe first recognised climate change as a looming environmental problem in 1985 by targeting scientific study of the issue in its research policy. Between 1985 and the joint Council declaration of 1990, the European Council agreed on a work programme aimed at developing specific targets and goals for greenhouse gas reduction. This programme was followed by a European Commission report in March 1990 advocating 'the urgent need for a clear commitment by industrialised countries to stabilise CO_2 emissions by the year 2000 and consideration of significant . . . reductions at the horizon 2010' (Skjaerseth 1994: 26–27). An expert group introduced the idea of fiscal measures (energy taxes) to accomplish these goals in the Spring of 1990 and in June of that year the European Council pressed for the earliest possible adoption of targets and strategies for emissions limits (ibid.: 27). European leadership in the Second World Climate Conference in November 1990 was assured by the ministerial declaration of 29 October 1990 that

> The European Community and member states assume that other leading countries undertake commitments along the lines mentioned above [i.e. stabilisation of CO_2 emissions by the year 2000 at 1990 levels] and, acknowledging the targets identified by a number of member states . . . are willing to take actions aimed at reaching stabilisation of the total CO_2 emissions by 2000 at the 1990 level in the Community as a whole.
>
> (Bergesen *et al.* 1994: 17)

An interservice group of representatives from the principal General Directorates of the Commission, led by Environment (XI), Energy (XVII) and Taxation (XXI) was formed to design an integrated climate change policy around taxes, subsidies, regulations, and monitoring mechanisms in anticipation of the United Nations Conference on Environment and Development (UNCED), to be convened in Rio de Janeiro in 1992.

Although industry hesitancy developed quickly over the role and nature of prospective energy taxation and regulation, in October and December 1991 European environment and energy ministers agreed to take action, including the introduction of environmental taxes, on climate change (Skjaerseth 1994: 28–32). Internal political wrangling in the subsequent months over the character of the tax did not prevent the European Community from reasserting in the run up to Rio the assertive stance it had taken in prior international meetings on Climate Change in Noordwick, the Netherlands, in November 1989 and at the White House Conference on Science and Economic Research Related to Global Change in April 1990 (Gerelli 1991: 171–179). Often working in close concert with green NGOs and the Group of 77 less developed nations and China against American resistance, Community and member state delegations to UNCED were instrumental in drafting a Framework Convention that emphasised familiar EC ideas including (soft) emissions targets, differentiated responsibility among developing and industrialised parties, and burden sharing by wealthy nations through financial transfers to compensate developing societies for the costs imposed by measures to mitigate climate change.

111

Especially following the completion at Maastricht of the Treaty on European Union (TEU), Europe stood on firm legal ground in putting forward a bona fide claim to environmental leadership at Rio. No comparable OECD political entity has a constitutional base whose environmental credentials are as solid as those of the EU (Dinnage and Murphy 1996). The concordance between these constitutional provisions of the EU and leading doctrines of international environmental law was manifest in the European vision of the international regime for climate change inaugurated at Rio. First, the FCCC recognises at the beginning of its commitments clauses the principle of common but differentiated responsibility for climate change problems. Article 4.1 imposes on all parties to the FCCC reporting and monitoring obligations on their greenhouse gas (GHG) emissions. But Article 4.2 places the sole obligation to mitigate emissions on the developed economies which have historically been the major sources of rising atmospheric concentrations. Articles 4.2(a) and (b) impose very soft or non-mandatory aims on the industrialised states to return by the end of the present decade to their 1990 emission levels. No targets, soft or binding, were agreed for the period beyond the year 2000.

Second, the principle of burden sharing is incorporated in Article 11 which describes a financial mechanism or fund to which developed economy parties should make (undefined) contributions to support the incremental costs of environmental responsibilities that may be undertaken by developing nations. Third, the FCCC makes only ambiguous reference to international trading in emissions rights. Article 4.2(b) mentions that parties may seek to return to 1990 emissions levels, 'individually or jointly', but leaves for later negotiation in a separate protocol what the legitimate role of trading might be. Even though both Germany and Norway had earlier advocated joint implementation, the EU did not bargain hard for a clear endorsement of JI sought by the United States and opposed by G 77 on the grounds that each polluting nation or regional organisation should deal with its emissions at the source (Sydnes 1996: 202, 279).

Finally, under the authority of EC Treaty articles 130s(1) and 228(3), the European Union signed the FCCC as a party in its own right, as did each of its member states.[2] Article 22(2) of the Convention states that where both regional organisations and one or more of their constituent member states is a party to the Convention, 'the organisation and its member States shall decide on their respective responsibilities for the performance of their obligations under the Convention' (ILM 1992: 870–871). This double adherence recognised the collective power of the Union as an international actor, but left unspecified how the shared obligations of compliance might be allocated among the member states. The relatively less developed cohesion bloc states within the EU (Spain, Portugal, Ireland and Greece) formally assumed individually the soft legal aims taken on by all developed nation signatories when they ratified the FCCC in their national capacities.

Yet, they have suggested that their agreement was implicitly conditioned on the EC meeting its stabilisation goal as a regional entity with internally differentiated responsibilities that would allow their emissions growth to be offset by emissions reductions in the richer member states. European practice since ratification has reflected this understanding, but the direction and coordination of member state obligations and financial transfers in a common policy that would aspire to the global leadership on climate change proffered to UNCED has been far less forthcoming.

A Community programme for climate change?

Consistent with the two track ratification process, EU climate change policy has followed a double path that is both national and regional. On the national track, each member state has adopted an internal soft target or, at the least, an expected trajectory for CO_2 emissions.[3] Four states declared reduction goals. Belgium and Denmark announced 5 per cent reductions from 1990 levels by 2000. The Netherlands pledged a 3–5 per cent cut from 1989/90 by 2000; Germany a 25–30 per cent reduction from 1987 levels by 2005. Three EU-12 nations (the United Kingdom, Luxembourg and Italy) agreed to stabilise emissions at 1990 quantities by 2000. The stabilisation objective was also adopted by the new EU member states of Sweden and Austria. Finally, six member states indicated they would have to increase emissions in 2000 over the 1990 baseline. France set a per capita emissions target that effectively led to an expected 13 per cent (now revised down to 7 per cent) expansion. Ireland, Greece, Spain and Portugal reported that their energy needs for economic development would require, respectively, 20 per cent, 25 per cent (later revised down to 15 per cent ±3 per cent), 25 per cent (later revised down to an 11–13 per cent increase) and 30–40 per cent emissions increases.[4] Finland only hoped to stop the growth of CO_2 emissions by the end of the decade (projected as an increase of 16 per cent over 1990 levels), due to its perceived need to enhance energy security by substituting domestic electricity, with a heavier consequent reliance on fossil fuels, in its base load production for existing imports (EC 1996: 32–33). If it is assumed that each of the member states were to reach its stated goal, the ability of the EU as a regional organisation to meet the FCCC stabilisation objective will depend principally on a German reduction of some 12 per cent of national emissions by 2000, that would serve as the major counterbalance to expected increases in the other EU member states.[5]

The ability of member states to comply with targets or trajectories depends primarily on three variables. The first is the economic context in which emissions mitigation proceeds. Other things being equal, economic and population growth stimulate energy and transportation demand and emissions. Conversely, industrial transition to service economies and industrial restructuring or replacement of obsolescing capital plant tends to reduce fossil fuel intensity.[6] A second influence on emissions arises from changes in energy markets and policies not tied to climate change measures *per se*.[7] Beyond economic and energy factors that form the context in which climate change analysis is framed, emissions levels are evidently a function of dedicated policies and measures imposed by national and regional governments or by international treaties.

European nations have focused on four policy areas in their dedicated climate change programmes:

1   efficiency gains in the generation and transmission of energy;
2   efficiencies in the consumption of energy in the manufacturing, commercial and residential sectors;
3   fuel switching toward less carbon intensive sources;
4   reductions in emissions from the transportation sector through greater automobile engine efficiency and the extension of mass transportation infrastructures.

National communications indicate that member state programmes differ about which of these areas will be stressed and in what proportions they will be pursued through regulations, taxes, subsidies or voluntary initiatives.[8]

In the aggregate, national programmes do not rely on the extensive use of mandatory regulations or economic instruments as key measures by which targets will be implemented. No member state has proposed emissions permit trading at the national level. Sweden, Denmark, Finland and Holland (and Norway outside the EU) have enacted CO_2 taxes, but they are characterised by very low rates and wide exemptions for energy intensive industries fearful of the impact of such taxes on their international competitiveness. Instead, national plans have usually given primary emphasis to so-called 'no regrets' options and business–government cooperation in their implementation. Strategic reliance on no regrets solutions suggests that there is an extensive portfolio of GHG emissions-reducing options, in production and consumption, unexploited by normal market operations, which can be installed and operated at no net cost to their users.

The idea of no regrets could be meaningful in both a social and a private context. In a collective sense, no regrets could signal that a change in policy could increase societal wealth while emissions fall, with a consequent improvement in public welfare. This 'double dividend' would couple environmental improvement with positive employment and growth effects.[9] However, the theoretical simplicity of no regrets definition vanishes as soon as we acknowledge that policies in place have been enacted, and will be defended, by entrenched political forces. Real transactional costs of political mobilisation are associated with displacing embedded policies and, in practice, the resulting political opposition has shrunk the pool of accessible low (or no) cost mitigation options. As the total prospective costs of national

CO_2 mitigation targets have climbed higher than proponents of aggressive climate change action would have predicted or hoped, the probabilities of compliance have correspondingly declined.

The number of no regrets options in the private sector is equally elusive (Heller 1996: 317–320). It is often asserted that firms or consumers do not invest in new technologies or the development of products that yield a competitive rate of return and are ecologically more beneficial, due to positive costs of information search or organisational or financial reform. Unless these costs are offset through some public sector subsidy, businessmen and consumers, who do not distinguish transaction costs from production costs, will resist the imposition of regulations to force the adoption of these solutions. If member state programmes were launched on the expectation of a broad putative stock of no regrets options and that stock disappears in the flux of economic and political practice, national goals reliant on the voluntary acceptance of emissions reductions may again tend to fall short of compliance.

The Community climate change programme was formulated in response to complementary hypotheses that portended insufficient national actions to meet CO_2 targets. First, national programmes with varied policy mixes could yield differential economic burdens on industrial sectors.[10] Unless climate change measures were truly no regret and without distributive effects, the alternative national patterns of taxes, regulatory costs and subsidies could distort competition within the European market. Since the major thrust of Community governance has been the elimination of national measures which affect internal competition, it was logical that the Commission would see a substantially harmonised EU-wide climate change policy as the preferred policy outcome. Second, in the run up to Rio, it was the EC which staked the European claim to moral leadership in the creation of a global climate change regime. The normative commitment to burden sharing, which Europe supported in the FCCC, also applied in microcosm to the less and more developed sub-regions of the Community. Unless a unified European policy could be defined that reflected an intra-EC agreement on how the costs of emissions mitigation would be distributed within Europe, it would be problematic whether member states would comply with the soft or non-binding objectives they had postulated in the absence of relative certainty about the behaviour of their peers. In this situation, Community institutions assumed a dual role in managing a negotiation about the internal distribution of obligations to mitigate and associate transfer payments, and coordinating or monitoring compliance with the EC accord.

Brussels projected in 1992 that, in the absence of new stabilisation measures, Community CO_2 emissions could increase by at least 12 per cent between 1990 and 2000, and described a regional plan to close that gap.[11] Although the Commission affirmed its adherence to a CO_2 strategy balancing the principles of efficiency, equity and subsidiarity, the major stress was on the cost effectiveness of the plan (EC 1992c: 165–166). In part, cost minimisation was reflected in an emphasis on no regrets measures to promote the rational use of energy in the Community's SAVE programme (EC 1992c: 32–53). Even more, economic considerations lay behind the adoption of a European energy tax as the centrepiece of the programme. Prior analysis had indicated that the marginal costs of emissions mitigation varied substantially among member states (Barnett 1992: 3–24, European Commission 1992c: 31). If national targets were assigned according to a politically set formula and transboundary action to meet those targets was not legalised through emissions trading, it would not be the case that only the lowest cost sub-set of mitigation projects would be realised. Conversely, while a uniform European energy tax would ensure cost minimisation, the burden of paying for climate change might be oddly distributed. In opting for a Community tax, the Commission implicitly committed itself to some unspecified system of transfer payments that would compensate those member states whose share of the overall, if minimised, costs of mitigations would be deemed inequitable.

The Commission's proposed climate change plan was composed of five measures, whose combined impact on emissions was projected to eliminate the 12 per cent EC shortfall below stabilisation. A 1.5 per cent reduction was to come from several new activities to be added to the existing JOULE and THERMIE programmes that focused on energy efficiency standards and research support in the areas of household appliances and transportation.[12] Further gains of 3 per cent and 1 per cent were expected from the new SAVE and ALTENER initiatives. SAVE was put forward as a series of no regrets regulations intended to require or induce private actors to behaviour which they would discover to have lowered their costs of operation (EC 1991, 1992d). The initiative included measures on building certification, energy billing, energy service companies, energy audits, third party financing of investments in efficient technologies, and mandatory inspections of boilers and motor vehicles. ALTENER offered EC subsidies for research and development in renewable energy sources. The correction of the largest share (6.5 per cent) of the stabilisation gap was assigned to the carbon–energy tax to be implemented at the Community level. The final component of the Community programme was the installation of an EC monitoring mechanism to which national climate change programmes must make annual compliance reports.[13] The monitoring mechanism was charged with developing a uniform data reporting system, common methodology for national inventories of greenhouse gas emissions, and a duty to evaluate independently national communications.

The history of the EC climate change programme has been less happy than its origins. ALTENER research in renewables was funded at only 40 million ECU for a five-year period. The SAVE initiative ran into varied degrees of industrial resistance that resulted in its transformation from a portfolio of mandatory regulations to a list of exhortations which member states might or might not enact as national legislation. But the carbon–energy tax soon emerged as the serious battlefield on which the politics of EU climate change would be decided. Opinion about the tax proposed by the Commission was always uneven among the General Directorates concerned. Support was lukewarm in DG II and DG XXI concerned with economic and fiscal policy (Skjaerseth 1994, Liberatore 1995). Because the EC treaty basis for the energy tax (130s(2)) demanded unanimity of all member states, French arguments that carbon alone should constitute the tax base and the British claim that subsidiarity considerations should preclude European level taxation were especially salient.[14] Legal issues were unsettled about whether international trade regimes would deem such a tax rebatable on exports or chargeable on the energy content of imports.[15] From a broad range of interests, the tax was barraged with criticism as, first, ineffective environmental policy because of the leakage associated with partial international regimes;[16] second, regressive in its impact on poorer EU member states and income groups; and third, economically damaging to the competitiveness of European industry, if enacted unilaterally.[17]

The controversies about the distributive and competitiveness effects of environmental taxes led to a series of compromises that defined the contours of the final Commission submission. First, the tax base was divided between a pure carbon tax and a more general energy tax. One-half of the proposed tax was to be levied on the CO_2 content of a fuel's emissions; one-half collected from an equal rate charged on all fuels, regardless of carbon content, except for renewables and some hydropower (EC 1992e: Article 8, 5 and Article 3, 3–4). Next, the tax was to be phased in over the stabilisation period to allow adjustment. The total rate of ECU 17.70 per ton of oil equivalent corresponded to $3 per barrel (1991 prices in 1993) and was scheduled to rise to $10 per barrel equivalent by 2000.[18] Third, consistent with the subsidiarity claim, the tax was to be collected and retained by national governments and not the EU. Fourth, to forestall industry fears of increasing overall tax burdens in a recessionary economy, the tax was required to be fiscally neutral or offset by decreases in other national taxes (EC 1992e: Article 15.2, 7).

The imposition of neutrality allowed the Commission to argue that the energy tax could be a no regrets measure and pro-growth, since it could displace arguably more inefficient tax burdens on labour or social security about which European business frequently complained. However, neutrality and national tax retention also had their costs. The Community would not have access to the revenues of the energy tax itself to manage burden sharing among the member states. Other sources of transfer would be needed to counteract any inequitable effects of an efficient energy tax. In addition, tax revenues would not be available for international transfers in conjunction with Article 11 of the FCCC or with EU expressions of desire to subsidise the rebuilding of the energy infrastructure in the transiting states of Eastern Europe (EC 1993: 19–20).

Three added provisions built into the carbon–energy tax attested to the intensity of the concerns over the impact on EU competitiveness of unilateral climate change. To protect energy intensive industrial sectors from imports from third countries that had not introduced similar taxes, the Commission could authorise member states to grant graduated reductions in the tax payable by firms whose total energy costs exceeded 8 per cent of total value added.[19] Further compensating exemptions or tax refunds were legislated for new firm investment in energy efficient equipment (EC 1992e: Article 11, 6). Finally, the entire tax proposal was prefaced by a conditionality clause that deferred the applicability of the tax until other member countries of the OECD introduced a similar tax or other measures of equivalent financial impact (EC 1992e: Article 1.2, 3). Acceptance of the conditionality requirement conceded capitulation by the advocates of a leading role for Europe in climate change politics. Nevertheless, the exemption provisions did not allay worries about prospective sectoral losses in traded energy intensive goods and services or inter-industry impacts on automobile firms. Additional resistance attached to the fears of the poorer Community regions over the distributive problems of uncertain burden sharing.

Given the combined force of this opposition, the carbon–energy tax proposal lay stalemated in political limbo until 1995 when the Commission made an amended submission which removed the compulsory character of the levy (EC 1995). Member states were urged to pass uniform taxes similar in form to the discarded mandatory tax with the rates left up to the national governments during a transitional period to harmonised rates that would end on 31 December 1999. The conditionality clause could be deleted from the amended proposal since no member state was required to do anything. In the wake of the evisceration of the force of the SAVE initiative and the low appropriation for ALTENER, the abandonment of the mandatory EU level tax returned, by default, climate change policies to the member states. In spite of compelling single market arguments for a unitary European regime, the Community role has been rolled back to the monitoring of national programmes whose fulfilment could at most be urged or criticised.

Three hypotheses, in their separate or aggregate effects, might account for the failure to enact a regional climate change policy for the EU. The first would credit the opposition of industrial sectors facing economic losses from the relative price effects of climate change measures. The second hypothesis would suggest that coordination problems associated with incomplete regimes and leakage outside of the Union underlay the blockage. Finally, even if climate change measures effectively reduced EU emissions without causing offsetting foreign CO_2 growth, the costs of mitigation in Europe could be deemed too expensive compared with the damages to Europe that climate change threatens. To the extent that all three of these hypotheses may have contributed to the vulnerability of the proposed EU programme, a revised approach to stabilisation aims, either through national programmes alone or through a European level emissions trading system, must offer a serious prospect of reducing one or more of these categories of asserted economic costs.

The prospects of subsidiarity

With EU climate change policy operating principally through the monitoring mechanism and the Community delegations to the several bodies of the FCCC regime, the member states have inherited the task of implementing national policy measures to comply with the European stabilisation target. It was the heavy weight of this burden which had led the Commission in 1992 to predict the 12 per cent shortfall in emissions reductions in the absence of EU additional action. Even after the submission in 1996 of the most current data on national emissions trajectories and programmes in the Second Evaluation Report, it remains difficult to assess the accuracy of the earlier forecast. The report estimates that the EU-15 as a whole will just achieve its stabilisation target (EC 1996: 23–24). This contrasts with a projected trajectory of a net 14 per cent gain in CO_2 that would have occurred in the absence of mitigation measures. The large emission declines are reported in Germany (77 per cent) and the United Kingdom (20 per cent), which together make up 97 per cent of the gross prospective EU reduction that offsets emissions increases in nine other member states.

The Commission confesses to serious scepticism about these highly asymmetrical numbers (EC 1996: 14–20; Marchetti 1996: 298–327). The contingency of aggregations of national trajectories is underlined by the EU monitoring mechanism which, in modelling EU-15 emissions under slightly changed economic growth and fuel price assumptions, estimates a likely 3–10 per cent overshoot of the 2000 target (Climate Action Network and Climate Network Europe 1995, EC 1996: 18–19, 25–26). Even assuming the currently reported aggregate national CO_2 trajectories are accurate, the prospects for future effectiveness of a European climate change politics consistent with the principle of subsidiarity that calls for action at the level of the EU member states remain problematic. Analyses of the early national experiences in this field suggest several reasons for caution. Studies of the internal politics of key member states argue that there is neither public salience about climate change issues nor institutional mobilisation of technical or interest group constituencies sufficient to produce measures that threaten economic growth or accepted life styles. The result of this imbalance has been a failure to enact and implement national policies that are aimed at climate change *per se*. The factors that have led to emissions declines in some European nations and lowered emission growth trajectories in others are better attributed to contextual variables or to the secondary effects of measures originating in policy agendas unrelated to climate change. Whether or not the inability to pass new measures compromises EU compliance with the 2000 stabilisation target, it bodes ill for the post-2000 period when the dampening effects on emissions growth of important contextual variables dry up and CO_2 trajectories again begin to rise.

The hypothesis that most emissions gains that have been registered in the Second Evaluation Report stem from ancillary effects of measures unmotivated by climate change politics, can be illustrated in the cases of the United Kingdom and Germany, the two nations whose emissions reductions provide almost all of the EU's gains. The more than 6 per cent reduction projected for UK emissions results principally from fuel switching associated with the privatisation and deregulation of the domestic energy industry and the termination of subsidies to the coal industry.[20] The German case provides particular cause for concern, given Germany's pivotal role in EU compliance with FCCC targets (Bundesministerium für Umwelt 1996, Sydnes 1996: 194–195, 200, 204–205). German CO_2 emissions fell by 12.7 per cent between 1990 and 1995. However, although over this period per capita emissions in the former West Germany fell about 3 per cent, population and economic growth in the old Bundesländer led to a 2 per cent increase in absolute emissions. By contrast, CO_2 decline in the new states of the East was more than 46 per cent, due to a population decrease of 4 per cent, import displacement of local production, falling consumption of lignite, and improvements in energy efficiency tied to the upgrading of plant and infrastructure capital to Western European standards. With rising economic growth in the former GDR, incremental carbon savings there have fallen off from 24 per cent in 1991 to 5 per cent in 1993 and 3 per cent in 1995. This exhaustion of the one-time effects of unification is reflected in the slackening of the pace of global German CO_2 reductions from the 12.7 per cent fall achieved between 1990 and 1995 and the predicted cumulative decline of only 13–15 per cent by 2000.

Like other European states, Germany has introduced no new taxes or broad gauged regulations in pursuit of a climate change agenda. There has been considerable promotion of the no regrets potential of a forward looking national programme of environmental regulations as a source of growth and employment through early adjustment to new technologies and market possibilities (Bundesministerium für Umwelt 1996: 212–221). Yet, the logic of no regrets has not had enough political force to eliminate coal subsidies long recognised as highly inefficient (Anderson 1995: 4–7). As a consequence, the burden of German climate change policy has come to rest squarely on the programme of voluntary industrial agreements announced in 1995–1996. This reliance is made explicitly apparent in the commitment of the Federal Government in the agreements 'to refrain from introducing additional regulatory climate-protection measures', 'to give priority to German industry's private initiative', and, should an EU-wide carbon–energy tax be enacted, 'to provide exemptions for industry participating in the voluntary commitment programme' (Bundesministerium für Umwelt 1996: 8). However, the effectiveness of voluntary industrial measures remains questionable without the threat of regulation or the appointment of an independent third party to monitor and report on implementation of the commitments.[21]

Data to permit realistic assessments of whether particular industrial sectors will be able to meet their commitments are not yet available. All sectoral projections and plans agglomerate mitigation actions in the old and new Bundesländer. Therefore, there is no attempt to retain the original German goals of a 25 per cent reduction in the former West Germany alone and still more extensive declines in the East. Although each sectoral agreement is idiosyncratic, the industry programmes of key emitters like cement manufacturers and electric power generators seem to illustrate typical patterns of expected reductions trajectories. As such, they provide little evidence to gainsay the argument that forecast reductions, if realised, will be caused mainly by unification and the business as usual replacement, motivated by high energy costs, of depreciating capital with more efficient plant.[22]

In summary, I suggest that the European commitment to leadership on climate change has become increasingly suspect in the years between 1987 and 1996. The failure to enact a Community-wide programme built around the carbon–energy tax has returned the onus of policy to member states unprepared or unwilling thus far to bear it. The unification of Germany brought into the European pool of mitigation options a substantial number of low cost, often subsidised, opportunities to limit carbon in the former GDR. It is principally the exploitation of this one-time pool by West German investors that has carried European stabilisation as far as it has gone. Against a background of rising emissions in all member states which foretells increasing difficulties in meeting FCCC targets in 2000 and beyond, we should note that nearly all of Europe's offsetting mitigation has come from an unforeseen institutional reform that created an implicit system of emissions trading outside the former boundaries of the EU. In this light, it is logical to ask whether the political economy of climate change can be improved by generalising the German solution.

Carbon trading and the European politics of climate change

Trading and taxes

Without recourse to a Community carbon–energy tax and without faith in the will or ability of member states to limit emissions, advocates of European leadership in climate change have renewed attention to emissions trading. The general idea of international markets in carbon reductions had been put forward by European delegations to the UNCED deliberations before Rio as measures to be incorporated in the Framework Convention. Germany, Norway and Italy have all claimed parentage of the initiatives that led to the inclusion of joint implementation in Article 4.2 of the FCCC.[23] Although subsequent multilateral negotiations have refused to credit offsets in JI trading between OECD and developing nations, small scale financing for demonstration AIJ has been provided by Norway, the Netherlands, Sweden and Germany.[24] More recently, the replacement of the abandoned Community carbon–energy tax with an intra-European emissions trading or JI market has been proposed to reanimate a regional climate change programme (Bergesen *et al.* 1994: 32–48; Koutstaal and Nentjes 1995). In distinction to trading between EU and developing nations, JI projects within the Community, or between Community states and nations in Eastern Europe, could be legally creditable under the Berlin Mandate that emerged from COP-1 in 1995.[25] Given the variations in the marginal costs of mitigating carbon emissions within this area, it would be economically rational to institutionalise a trading or offset system either within the Community or across the wider OSCE region. At the same time, we might also ask whether such a trading system is likely to avoid the distributive, coordination and total cost problems that contributed to the fall of the EU carbon–energy tax.

On efficiency grounds, it would be difficult to claim that trading could have rallied the political support denied to EU carbon–energy taxes. The strongest argument for taxes over trading is that taxes minimise the potential economic damage that environmental controls might produce.[26] The strongest argument in favour of trading over taxes is that with multiple, uncoordinated fiscal and regulatory policies affecting the price of energy, changes in other policies may offset the effects of the carbon–energy tax (Hoeller and Coppel 1992: 23–24). It becomes unclear how much protection environmental taxes will actually buy. Since the interests that defeated the carbon–energy tax feared economic losses far more than environmental ineffectiveness, it is unlikely they would have been mollified by the substitution of a trading system for the proposed tax. A *prima facie* comparison of the advantages of taxes or trading might suggest that taxes would be the preferred instrument on efficiency grounds, given Europe's relatively greater concern with economic growth than environmental protection.

Trading might be politically more feasible because of sectoral distributive effects. Although in theory the distributive effects could be made equal, in practice it may be easier to reduce the distributive impact of environmental controls by grandfathering. If permits are given to current emitters, they must bear only the marginal costs of the mitigation required, rather than a tax on all carbon released.[27] In addition, grandfathered permits may rein in the political temptation to see carbon taxes as a new and expandable revenue source. However, tradable permits would require that burden sharing between EU member states be built into the system by which the overall quantity of permits is allocated. The problem of distributing permits and compensating transfers could be as politically difficult as is inter-sectoral distribution for taxes. In the end, it may be hard to provide persuasive evidence that intra-European trading will be able to escape the political trap in which the carbon–energy tax lies enmeshed and inert.

The successful design of environmental instruments, whether taxes or trading, must consider the problem of the adaptation/mitigation threshold that may have been the determinative cause of the failure to enact the carbon–energy tax. Each nation facing the risks of climate change may choose a strategy of mitigating emissions or adapting to an altered environment. Mitigation demands collective action. Adaptation is a local response whose cost establishes a budget constraint for mitigation strategies. Because there are more and less expensive ways to mitigate, whether mitigation dominates adaptation for any given nation depends on the local adaptation cost, the mitigation options the collective regime makes available, and the confidence that is placed in the effectiveness of the collective solution. It is possible that EU states may not prefer adaptation to mitigation in all cases, but that adaptation does dominate the particular mitigation possibilities offered by either the carbon–energy tax or intra-European trading. To make use of the least cost pool of mitigation options within Europe may simply be more expensive than Europe's expected future cost of the risks of living with climate change. Even if, as the German solution indicates, intra-European trading is good economic policy because of the variation in the marginal costs of abatement within the region, the key to meeting the EU's announced goal of leadership in climate change may lie in expanding trading beyond Europe to include a still wider range of mitigation options.

The incorporation of emissions reductions opportunities with substantially lower abatement costs into an integrated market could drive the cost of mitigation below the putative threshold that makes adaptation the preferred political outcome. At present, such low cost opportunities are likely to be concentrated in Asia where new infrastructure investment in energy is demanded and the normal marginal efficiency of energy generation, transmission and distribution is low. In this view, the advantage of trading over taxes would lie less within Europe than in the access to broad international trading which could open this expanded set of cheap mitigation options. Assuming that the politics of multilateral negotiations will not allow the enactment of a global carbon tax and transfer system, EU strategy should make a double turn to break out of the impasse of European climate change policy. First, it should substitute emissions trading for environmental taxation as the heart of its initiative. Second, it should implement this initiative at the international, rather than the regional, level.

Reimagining the political economy of climate change

The Conference of the Parties of the FCCC is the obvious forum through which to establish a trading regime that opens to Europe an expanded opportunity set for low cost mitigation. The initiative undertaken at Rio reflected a widely shared optimism about the prospects for attacking problems of a global reach, with negotiations on a global scale. Yet the first years of the FCCC regime have been marked by an inability to resolve the core issues of how much wealthy nations must cut their GHG emissions; what commitments, if any, will be assumed by developing nations; and who will bear the costs of these mitigation activities. The secondary questions of climate change instruments, including JI and other trading systems, have been embroiled in the fundamental, intractable controversy among FCCC signatories over commitments and associated transfers (Sydnes 1996: 320–333). This deadlock is reflected in the decision by CoP-1 in the Berlin Mandate to establish a JI Pilot Programme that refuses any credit against national targets for JI investments between Annex I and developing country parties. I would argue that the FCCC process has stumbled over some problems that are specific to climate change and others that are peculiar to the UN multilateral forum in which climate change is addressed. Unless this stalemate is undone, either by a reimagination of the political economy of climate change in the CoP or by the removal of the regime building process to an alternative forum less burdened by the rules and culture of the FCCC, the promise of wide trading systems to reduce mitigation costs for EU and other OECD economies will remain unrealised.

Four premises, incorporated in the structure of the FCCC and the subsequent CoP negotiations, have contributed to the deadlock over climate change allocations and instruments. The first premise is reflected in the linked propositions of FCCC section 4.2 that developed countries must 'take the lead' in mitigating emissions and that differentiated responsibility exonerates developing countries, at least at the outset, from taking on commitments. Second, there is an assumption that Annex I parties have a substantial willingness to pay for mitigation and will agree both to finance mitigation on their home territories and contribute to the Article 11 international transfer fund that will subsidise environmental costs in the Third World. Third, the Convention specifies no schedule for mitigation action at the global level. At Rio and at Berlin, resistance by non-Annex 1 parties to universal commitments was pivotal to the negotiations, although it is acknowledged that most of the expected growth in GHG emissions will occur in developing countries in the coming decades. Fourth, there has been a sense among key non-Annex I nations that to allow OECD signatories access through trading systems to low cost mitigation opportunities would contravene their sovereignty and economic development. In part, non-Annex I nations have resisted trade because it lowers the price at which polluter states can 'escape' the punishing consequences of their past action. But there is also a perception that there is a fixed stock of cheaper mitigation possibilities in the developing world that constitutes 'low hanging fruit' to be preserved for later consumption at home.

The convergent effect of these assumptions leads toward a delaying strategy in regime negotiations on the theory that Annex I parties will eventually act in their own territories, agree to provide a multilateral transfer fund, and do so in time to avoid climate change damages so serious that local adaptive investments will have to be undertaken. This bargaining stance is particularly apt in the United Nations process where voting rules require substantial consensus and favour blocking coalitions and where the multilateral community has a historic preference for a regime that relies on national regulations and international transfers that is embedded in diplomatic culture.[28] However, progress in the direction of inclusive trading may depend on a new conception that challenges all of the premises that contribute to deadlocked FCCC negotiations. The demand for early, substantial and good faith action in the home territories of Annex I parties – a commitment at the heart of the stalled programme of EU leadership in climate change – calls for the most radical revision. The legal framing of environmental damage as an issue of historical cause and responsibility is ill fitted to a situation where the potential losses of climate change were unforeseen at the time when the long lived energy and transportation infrastructures of OECD nations were installed. Instead, an analytical focus on minimising the total costs of mitigation in a manner consistent with projected paths of economic development would push toward a concentration on initial action in the Third World, financed in part by First World investment. This conclusion follows from the proposition that it is less expensive, in general, to mitigate new, than to clean up or avoid existing, emissions.

Recent simulations, though controversial, suggest there are two likely substantial pools of relatively low cost mitigation options that should be the objects of a feasible climate change strategy (Richels *et al.* 1996). The first pool will lie in the advanced industrial economies at the time when new energy and transportation infrastructure is installed on a large scale. With respect to emissions from the production of energy, converting to electricity, fuel switching or installing late vintage infrastructure for efficient generation and transmission is far more economic than retrofitting capital in place. Similarly, the development of new transport systems that define rather than follow urban land use patterns promises lower marginal costs of mitigation than uprooting locked in capital stocks that are not yet fully depreciated (Ausubel and Marchetti 1996: 139; Nakicenovic 1996: 95–112). OECD energy capital stocks are now relatively stable, locked in the midst of a cycle in which the last round of infrastructure building anticipated the long-term growth in demand resulting from technological diffusion (Ausubel and Marchetti 1996: 143–151). Until over-capacity is used up and existing equipment depreciated, the scope for low incremental cost mitigation of emissions in these nations will be restricted. However, a climate change regime that is flexible enough to defer deep cuts in OECD emissions until these opportunities are ripe can save as much as 30 per cent of the estimated costs of more immediate emissions reductions (Richels *et al.* 1996: figure 4).

In contrast, in East and South Asia where demographic and economic growth combine for explosive demand and systemic under-capacity, there is a second pool of more currently available, inexpensive options to reduce potential emissions. If presently affordable mitigation is a necessary condition for superseding the climate change impasse, it is in the non-Annex I countries where the regime should direct and encourage early action. Mitigation opportunities in Asia focus on investment that upgrades the environmental quality of new energy and transport infrastructure, activities at the core of the development and transition processes where local capital is typically scarce. If JI projects or other conditional transfers from Annex I actors finance the incremental costs of improved environmental quality, projected rates of emissions increase can be restrained without dampening economic growth. By way of illustration, consider investment in the coal fired energy sector in China where the potential environmental benefits are large enough to be consequential for global climate change over the course of a 30–40-year capital cycle, but are available at low cost only when the capital cycle begins.

In 1994, China's total primary energy consumption derived 75 per cent from coal, 17.4 per cent from oil, 1.9 per cent from natural gas, 5.3 per cent from hydro and 0.4 per cent from nuclear energy (Yun 1996). Over the next ten years, China plans to increase its installed electricity capacity by 130–150 Gwe, of which 80–100 Gwe will be coal fired (International Energy Agency 1995a: 10, 1995b). While China's inevitable use of her coal stocks is a fundamental challenge for any global mitigation regime, CO_2 emissions from coal based power are far higher at present than they need be. Chinese installation of generation and transmission infrastructure, due to cost, organisational and regulatory constraints, is very often done either at low scale or, if at larger scale, with domestic rolling stock of an earlier technology vintage that reduces potential efficiency and requires more coal to be burned per unit of power output. The International Energy Agency concluded that the average net plant efficiency of World Bank financed projects (~35 per cent) is superior to domestically financed capacity additions (~30 per cent), but inferior to capacity additions in OECD countries (~40 per cent). State-of-the-art coal fired power stations can achieve net plant efficiencies of 43 per cent and more (International Energy Agency 1995a: 11). Small scale plants are less efficient still. The IEA suggested cost effective upgrades included increased average plant size, introduction of innovative and more energy efficient technologies to the power sector, and procurement of a larger share of equipment on the international market

126

(International Energy Agency 1995a: 12, 20–34; 1996: 24). JI support could reduce the capital costs, foreign exchange shortages and operational inexperience that have limited these possibilities. Similar opportunities for mitigation investments have been identified in capital upgrading, fuel switching and demand side management in the steel, cement, chemicals and transport sectors (World Bank 1994).

This pool of mitigation options evaporates as capital with a long lifetime is put in place. Worse, the development of an installed base of known technologies lowers the incremental costs of additions of similar stock. Exploitation of these one-time opportunities requires the design and implementation of a climate change regime that allows flexibility for Annex I organisations to invest where emissions control is most efficient. By itself, such 'where' flexibility may save as much as 70 per cent of the estimated costs of mitigation if mitigation options were to be restricted to Annex I nations. In combination with 'when' flexibility, the total costs of mitigation to higher, but probably sustainable concentrations of greenhouse gases has been estimated to decline by as much as 85 per cent (Richels *et al.* 1996: figure 4). Even if the magnitude of these savings and the policy mix needed to achieve them are contestable, the direction of the shift is not.

The importance for an effective climate change regime of locating and accessing the most affordable mitigation depends, in turn, on assumptions about the willingness of Annex I parties to pay for mitigation and the speed with which mitigation must occur to avoid predicted harms. The second problematic proposition that afflicts the FCCC process is the belief that Annex I states strategically conceal a high demand price for mitigation. No nation can be expected to pay more for a mitigation commitment than the price it would have to bear to adapt to climate change. Different nations will have varying threshold points between adaptation and mitigation that reflect their pool of low cost mitigation options and their potential economic vulnerability to climate change effects. A mis-estimation by negotiators of crossover points between adaptation and mitigation strategies may cause defections from a common regime. The likelihood that defection to adaptation as a default solution will become the dominant response by OECD nations is raised by the high degree of scientific and economic uncertainty about climate change. In addition, adaptation costs will only come due in the future under the watch of politicians not yet in office. Finally, nations may have serious doubts about the political will or legal capacity of other signatories in a collective regime to deliver on commitments which must be fulfilled if the benefits of their own mitigation measures are to be realised (Heller 1996: 312–315). If adaptation options impose a tight budget constraint on the cost of mitigation, bargaining that inhibits trading into the global pool of least cost options will miss the chance for agreement.

A third contestable premise of the FCCC process has been that holding out for delayed revelation of high willingness to pay is a viable negotiation strategy. The choice between mitigation and adaptation may have a time dimension that is politically salient. I argued above that fast closing time windows may offer unique opportunities to lower the cost of mitigation toward politically acceptable levels. Unless a sufficiently large number of nations commit to a climate change regime that allows access to the low cost pool within these windows, the receding horizon of affordable projects may lead to an inversely cresting preference for adaptation. In other words, if path dependencies are associated with the actions taken in emissions-intensive sectors like coal fired energy in China, enough warming may occur to cause some nations to take local steps to protect themselves against the effects of this change. In such a case, nations with a low adaptation threshold must agree to a mitigation regime before the behaviours that initiate path-dependent development occur. Deferred accord risks defections that will push the global system into an adaptive equilibrium. This contingent scenario might be called the 'China trap'.

It is recognised that emissions growth in the coming decades will be concentrated in the fast growing developing nations, especially those of East and South Asia (see Figure 5.1).

First, assume that much of the increase in China's emissions comes from its installation of an energy–transport infrastructure with a useful life of 30–50 years. The costs of retrofitting an installed base of private and public capital built on coal and the heavy use of automobiles can be, as experience in the United States shows, prohibitively high. Second, assume that global change models indicate that China's projected additional emissions from this installed base will produce a warming level 'X' unless China adopts, and is able to enforce, domestic regulations to internalise the environmental costs of its emissions. Third, assume that facing warming level X it becomes rational for a wealthy northern state to invest in a policy response 'Y' to

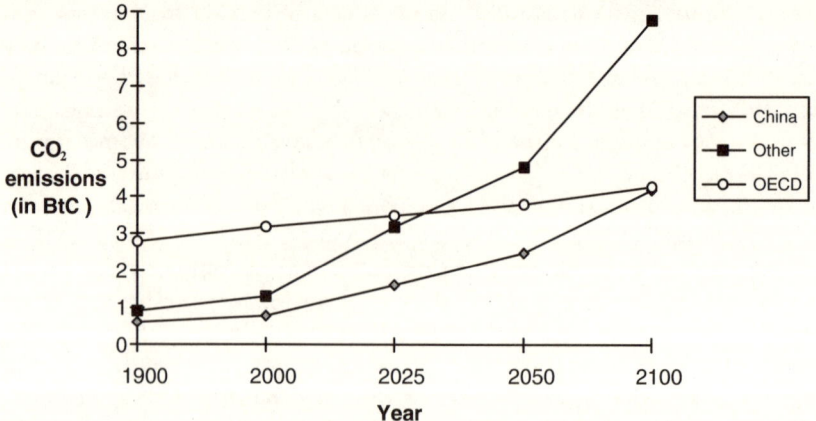

Figure 5.1 CO_2 emissions

adapt to the threatened damages from climate change. Fourth, assume that investment Y is lumpy, in the sense that once the investment has been made, it will manage damages over a wider range of climate change than the level X which induced the investment. Finally, note that once the adaptation expenditure has been made, prior investments in mitigation will be rendered useless. As long as actors are forced by warming level X to do Y, and Y is effective as a response over the entire temperature variation from present levels up to a defined level beyond X, it is irrational to spend any more than the cost of Y. This scenario implies that if industrial nations expect that additional emissions in China or other fast growing nations will, for whatever reason, reach certain levels, they may decide that adaptation strategies are their best option. An effective international regime must implement collective action to induce or require an alternative development infrastructure with lower emissions potential before the pursuit of business as usual springs the China trap.

The final assumption that troubles the FCCC negotiations, low hanging fruit, is also problematic. Ignoring time windows for low cost mitigation causes low hanging fruit to rot on the vine. It has been frequently argued that there is a discrete stock of 'low hanging fruit' which must be preserved for future mitigation responsibilities by the developing nations themselves. Although the more usual position of Third World states toward depletable resource stocks (e.g. oil) is that they are perfectly capable themselves of managing the trade-off between current international sale, domestic use and preservation for growth in asset value, the failure to credit JI investment in these nations during the pilot period is an effective bar to such sovereign choices. Whatever the putative justification for this policy, it makes little sense to treat as a depletable stock assets whose value is naturally wasting. Since many of the present options for fuel switching, energy technology upgrading or shifts in urban land use patterns will vanish with the installation of new capital infrastructure, there is no ability to bank these options for later local consumption. International trading of these wasting assets is the sole means of converting them to economic products.

If we reverse the FCCC–CoP standard premises concerning the site of early action, the high demand price and delayed timing for emissions reduction, and the bankability of mitigation options, the basic problem is to build a climate change system that allows those who can and will pay for affordable mitigation to have access to least cost stocks. Assuming that the pool in Annex I nations of short-term opportunities to upgrade and replace technologies and/or exploit no regrets solutions is limited, progress in regime building will turn on opening exchange markets with non-Annex I parties where low cost mitigation possibilities are concentrated. In this scenario, JI can be reconceived as the initial instrument for effecting that set of trades which could bring mitigation costs below the threshold of political resistance exposed in the EU carbon tax failure.

Systematic recognition of JI projects by Annex I nations in developing parties would automatically compromise issues that have stymied FCCC progress. To the extent that OECD parties relied on JI to comply with national commitments, it would become clear that the legal obligation to internalise the price of emissions does not entail a fine or tax equal to the cost of avoidance at the source. If the prospects of using international negotiations to impose punitive fines for historical emissions, to create a large reparations fund, or to cartelise successfully the supply of low cost mitigation options are abandoned as FCCC objectives, there is little reason for developing nations to oppose the growth of an affordable JI market. Like the worries over 'low hanging fruit', much of the political angst about JI expressed in the FCCC negotiations stems from a misplaced belief that trade in mitigation services is anti-developmental. As with other foreign investment, all JI projects require home country approval. There is then no risk of the unwanted sale of national non-wasting mitigation assets. In addition, because many investments in the incremental costs of cleaner energy and transportation will reduce SO_2 and NO_x as they lower CO_2 emissions, climate change action can assist with local and regional environmental problems far more salient in the Third World than global warming. Finally, because much of the initial focus of JI has been on carbon sequestration projects, developing countries have viewed it as either irrelevant to their priority development needs or as a threat to retard growth through 'carbon colonialism' (Heller 1995: 38). Instead, climate change and development goals should be mutually supporting. In an expanding JI market the greatest stock of low cost mitigation options is not in forestry, but in energy, transport and manufacturing infrastructure. Indeed, marketising emissions reductions could help to reduce the marginal cost of capital for large scale development projects that currently are bottle necked by a scarcity of concessional financing in emerging markets.[29]

The possible utility of JI as a central instrument in breaking the impasses that beset climate change politics in the EU and the FCCC is belied by the lack of private business activity in the JI pilot phase set up by the Berlin Mandate. Put most simply, in the absence of crediting for JI induced emissions reductions, JI will continue to be perceived as an innovative form of overseas development aid, rather than as a new field of foreign investment. However, the volume of either bilateral or multilateral assistance (e.g. the Global Environmental Facility), is far too low to dent the climate change problem. In many OECD nations, foreign aid budgets will decline still further. States have neither environmental technologies to be transferred as assistance nor the capacities to adapt them for sustained productive use in developing nations.[30]

The FCCC will provide no real test of the potential of an inclusive JI market until it substitutes a culture of business for the existing culture of aid.[31] Yet, the Berlin Mandate's pilot period for JI is rooted in the model of foreign assistance and, consequently, will fail to mobilise appreciable private resources. To generate inclusive international trading at a scale substantial enough to alter the emissions intensity of development patterns in rapidly emerging economies and, at the same time, lower their average marginal cost of mitigation to levels tolerable to their domestic politics, the EU and other Annex I parties must reposition their regime building strategy around an alternative approach to JI that escapes the particular politics which stall the negotiations of commitments and instruments in the FCCC.

Joint Implementation: beyond the illusion of Montreal

The structure of the FCCC and the CoP process that it initiated were inspired and shaped by the prior success of the Montreal Protocol. Although it has become apparent that important distinctions between the political economies of ozone depleting substances and greenhouse gases differentiate the problems that attend their control, analogues between the issues seemed to many to herald the dawn of a new era in international environmental law (Heller 1996: 302–303). As did the Montreal Protocol and its subsequent amendments, that led in less than a decade to an effective ban on CFCs, the climate change convention took the form of a framework agreement that was to presage an ongoing process of protocol negotiation. Because, like ozone depletion, the problem to be addressed was a collective good that implicated the entire global community, the natural forum for regime building was the United Nations, and the appropriate decision rule was diplomatic consensus among sovereign nation states. Finally, although there has been little trading under the Montreal regime, the legal institution is a cap and trade system with a supplementary transfer fund to compensate Third World nations for incremental compliance costs they might incur. In Montreal emissions permits, tied to an increasingly tight phaseout schedule, were grandfathered among the few OECD nations producing CFCs in any quantity. At the formal level, the legitimacy of cap and trade as the ideal model to be emulated in future inclusive multilateral negotiations was reified. At the substantive level, the pervasive differences in the interests threatened by limitation of emissions of GHGs and CFCs have rendered that ideal unrealisable in FCCC practice.

The fundamental deadlock in the FCCC process stems from failure to reach consensus on a universal allocation of caps or emission allocations that must, in standard theory, precede the initiation of trade. Developing nations resist commitments on the grounds of historical injustice and inadequate transfer payments. Annex I nations judge the costs of deep immediate cuts at home excessive and question the effectiveness of mitigation plans that do not comprehend emissions growth in all major parties. It has been easy to form blocking coalitions under consensual rules, especially in the presence of bad faith players (e.g. oil producing states) who are parties to the

Convention. A solution to this impasse that prevents the emergence of a trading regime must involve a change in the rules and expectations of the process by which the structure of the system is defined. This shift could occur within or outside the FCCC negotiations. In either case what is essential is the decision to abandon the quest for an inclusive agreement on caps as a prior condition to widespread emissions trading through JI.

In place of the multilateral cap and trade model, we can outline the development of an alternative international regime building process that incorporates least cost mitigation opportunities, achieved in part by an emissions market that does not require widespread consensual acceptance of national quotas before trading begins.

1   A committed core of Annex I nations would agree on a sustainable, stabilised level of greenhouse gas concentrations (e.g. 550 parts per million) by the year 2100. The stabilisation goal need not require level emissions over the next century, but does demand that a time path be specified whose average flows will yield the target stock.

2   Each core nation would agree to fixed limits on its own emissions, consistent with the stabilisation target, taking into account expected emissions, as influenced by business as usual and regime policy, from nations not currently accepting mitigation obligations. Accord on a global stabilisation target and national commitments is more likely among Annex I states than among all FCCC parties because they can negotiate their actual willingness to pay for mitigation, and design a regime that reflects that demand price, in the absence of strategic blocking behaviour by nations not participating in this minilateral agreement.

Rules for compliance with national commitments would allow for both 'where' and 'when' flexibility. 'Where' flexibility implies that trading affords the possibility of access to least cost mitigations options outside the core nations. 'When' flexibility allows mitigation action to follow optimal timing patterns across the twenty-first century. Mitigation would initially focus on reducing the rate of growth in emissions through JI in the developing world where infrastructure now expands. The focus would later shift to deeper reduction options that will arise in Annex I nations as large scale replacement of the depreciated installed capital base quickens. Appropriate benchmarking of emissions performance in OECD states against their optimal mitigation time paths, as well as research and development policies and measures to ensure that cost-effective technologies become available in phase with capital cycles, should complement the early action trading system.

3   Core states must establish national incentives to induce private actors to behave so as to comply with national commitments. These may include voluntary industry agreements, subsidies, regulations, taxes or tradable

permits. Whatever portfolio of instruments is enacted, credit against national liabilities or eligibility for subsidies should be allowed for qualified investment that reduces mitigation below expected baselines in nations which have not assumed commitments in the minilateral regime.

4   Because states outside the core will not accept national emissions quotas in multilateral negotiations, a functional equivalent for caps must be devised to move toward a comprehensive and effective regime. A competitive, multi-local process for setting project and policy baselines can be substituted for the political and centralised determination of allocations in a stalled FCCC. Project baselines limit trading credits to investments that would not have taken place in the absence of the offset mechanism. Policy baselines are surrogates for national quotas in that they represent commitments, made by national or sub-national polities, either to remove existing subsidies and other price distortions or to enact positive policies that raise domestic environmental quality above a business as usual projection of emissions.[32] Core nations can steer investment toward polities willing to assume such surrogate commitments by variably discounting the credits they offer for JI projects in those jurisdictions that bid in more environmentally ambitious policy baselines. Credit would be measured by the difference between emissions after investment and expected emissions under the policy baseline.

5   The minilateral JI system should be administered at the national level. In part, multilocality may be necessary because the same political tactics that frustrate universal caps in the FCCC will impede multilateral agreement on appropriate JI qualification standards. In part, decentralisation will enhance learning and an evolution toward viable common standards to ensure the system converges on the stabilisation target. The variables that could enter the definition of project and policy baselines are multiple. What private actors would do in the hypothetical world of expected trajectories of baselines as usual is speculative. Moreover, if JI is to increase mitigation, there must be assumptions about the regulatory context which would have affected the volume of emissions in JI's absence. In the face of such complexity, it is unlikely that any predetermined uniform legal standard for JI qualification will be as efficient, at the outset of a regime, as would be allowing core participants to charter diverse national or regional processes of certification that set in motion a dynamic game. Competition within a multi-site, coordinated game could produce a positive environmental spiral. Conversely, participants in a minilateral core agreement will need a coordination mechanism to reduce temptations for nations home to JI investors to legislate lax qualification rules. Ongoing monitoring and intra-core negotiation of national standards to define emissions baselines can prevent a race to the bottom in pursuit of cheap projects that do not reduce emissions below what would

have happened without JI, incentives to seller nations to lower environmental quality to have more JI projects to market, and a trade war that produces a general deterioration of global environmental quality.[33]

6   The final aspect of an evolutionary JI process that leads to an inclusive international trading regime is convergence around what the market reveals as the actual demand and supply prices of climate change mitigation. Core governments that decide whether to credit proposed JI projects can induce key developing nations to participate in the regime by negotiating with them differentiated policy baselines which reflect their strategic importance to the international system. Competition for limited capital stocks should attract other sellers of mitigation services whose bids for shares of available funds will be represented in offers to assume national baselines that define their environmental commitments. As the circle of nations drawn into the regime widens, the benefits for holdouts to act as attractors for GHG intensive activity will increase with the scarcity of leakage possibilities. At some point it may become necessary to enact trade sanctions against such holdouts. However, both the political effectiveness and legal legitimacy of such actions grow with the number of participants in the minilateral regime (ILM 1991: 1620).

A successfully managed back-door approach to a comprehensive JI regime will result in an expanding number of nations seeking to enter the game. Developing nations will join to realise the monetised value of their comparative advantage in the provision of mitigation services. Annex I states will play to compensate their real demand for lowering the risks that climate change threatens. In the end, such a dynamic process can lead to a behaviourally expressed accord on acceptable levels of mitigation costs and transfers, now hidden by strategic politics in the multilateral forum. A progressive revelation by the market of realistic solutions to the climate change problem may interact with the FCCC negotiations in several ways. On one hand, the presence of a competitive forum to define a climate change regime may cause coalitions in the multilateral process to abandon the bargaining strategies that now produce deadlock, so that the diplomatic process may reclaim an exclusive jurisdiction for the FCCC. On the other, the FCCC might simply await an evolved solution in the parallel minilateral forum which will then be represented in a treaty as the natural outcome of the multilateral process. Here, we might surmise that the legal celebration of a comprehensive multilateral regime should be reimagined as the last, rather than the first, step in building international institutions.

Conclusions

Four principal findings and arguments emerge from this study:

1   The sincere effort of the European Union to assume leadership in the politics of international climate change failed to result in the passage of an effective EU regime for intra-European emissions mitigation. The key EU initiative to enact a carbon–energy tax stalled principally over fears about its excessive costs for economic growth and competitiveness and a secondary concern about its consistency with the EU norm of subsidiarity.

2   National programmes by EU member states to meet regional and national FCCC stabilisation commitments are unlikely to be successful in the absence of joint agreement on coordination and burden sharing.

3   By itself, a shift in focus from taxation to intra-European emissions trading does not resolve the cost problem implicit in the failed tax campaign. To move toward political viability in Europe, it would be wiser to extend international trading to offer EU actors access to a much wider pool of low cost mitigation options in developing countries.

4   Deadlock in the FCCC negotiations over national commitments prevents agreement on an inclusive regime for Joint Implementation, a practical form of international emissions trading. Europe can reassert a leading role in escaping this dilemma by pursuing JI as a central element of a long-term climate change policy, emphasising early action in rapidly growing Asian economies and greater reliance on competitive markets to reveal the value of carbon mitigation.

Notes

1 See ILM (1992: Article 4.2). For a brief introduction to Joint Implementation, see Heller (1996: 297–301). The First Conference of the Parties (CoP-1) of the FCCC endorsed a pilot phase for a type of JI, labelled Activities Implemented Jointly (AIJ), although the utility of this pilot phase is problematic. See ILM (1995: Decision 5/CP.1).

2 The EC ratified the Framework Convention on 21 December 1993. With Belgian ratification in 1996, member states completed their individual ratifications.

3 See Bergesen et al. (1994: 23). There is extreme variation in the form and completeness of the national communications of the member states and in the substantive nature of their commitments. Greece, Ireland and Portugal would assume no official CO_2 targets, consistent with their position that European objectives should be calculated and met at the regional level alone. France expressed a preference 'for commitments on policies and measures rather than to any quantified emissions limitations' and would agree only to a general standard of maintaining per capita emissions under 2 tonnes. Given population growth and the fact that current per capita emissions fall somewhat below this standard, stabilisation is not likely. Germany elected a target time line (2005) that was idiosyncratic. The history of national commitments and objectives, as well as the best current estimates of emissions trajectories, are reported in European Commission (1996: 2–3, 6–7, 23–24, 27–45).

4 In these instances of downward revisions of growing emissions trajectories, the major influence has been lower economic growth in the early years of the 1990s. See European Commission (1996: 37, 43).

5 Apart from targets and originally forecast trajectories, revisions based on the most current figures have been included in the Second Report of the Monitoring Mechanism. Increased emissions have been projected in Portugal (36 per cent), Finland (29.7 per cent), Spain (23.8 per cent), Ireland (20.5 per cent), Greece (14.3 per cent), France (8.7 per cent), Sweden (4.4 per cent), Italy (2.9 per cent), and Austria (0.6 per cent). Reductions were projected in Luxembourg (24.1 per cent), Germany (13 per cent), Denmark (11.9 per cent), the UK (6.1 per cent), Belgium (1.1 per cent) and the Netherlands (0.4 per cent).

6 See generally for sector by sector plans in this regard, German Ministry of the Environment *et al.* (1996).

7 The most obvious example of emissions-reducing policies not connected to climate change is France's long standing commitment to nuclear energy development. Similarly, Spain has pushed the substitution of natural gas for dependence on foreign oil supplies. However, the independence of such policies from climate change considerations is illustrated by the push in Finland to reduce its reliance on imported energy and the planned Swedish phaseout of nuclear power, for distinct environmental reasons, after 2000, even though, in each case, CO_2 emissions will thereby be increased. See European Commission (1996: 33, 43–44).

8 See European Commission (1996: 27–45) for specific measures proposed in each member state.

9 See discussion of no regrets options and citations in Heller (1996: 315–316). See also Anderson (1995), Sydnes (1996: 212), European Commission (1992a: 9). There has been substantial debate over whether there can be a 'double dividend' from energy taxation. See generally, European Commission (1992b).

10 The implications of distributive effects for strategic environmental policy, and the legal limits of devising an industrial or offset policy to combat them, are discussed in the introductory chapter of this volume.

11 See European Commission (1992a: 5–6). It is not clear from the documented plan which factors influencing CO_2 trajectories the Commission took into account in reaching its 12 per cent shortfall estimate. See European Commission (1996: 25).

12 These estimates and those for SAVE, ALTENER and the Community energy tax, discussed below, are given in European Commission (1992a: 5–6).

13 COM92(181), adopted as Council Decision 93/389/EEC.

14 France's case was built both on the efficiency effects of linking the measure of the tax directly to the emissions volume fuels and the favourable competitiveness effects stemming from France's earlier conversion to nuclear energy. The distinction between unanimous and qualified majority voting is also discussed in Chapters 2, 3 and 4.

15 No hard legal determinations seem to have been reached on whether the import of goods from nations without comparable taxes would be countervailable as a subsidy. For consideration of the range of these issues, see Düerkop (1994).

16 The problem of leakage is not specific to the proposed carbon–energy tax, but to any incomplete regime in which costs are imposed non-uniformly. Leakage may occur in theory mainly because of price effects on fuels or incentives for energy intensive industry to relocate. See Smith (1995).

17 A large literature which varied widely in its analytical questions and conclusions was generated by the controversies about the carbon–energy tax. See for example, Welsch (1995a, 1995b); Denis and Koopman (1995); see also the five analyses done for the European Commission of the efficiency and distributive effects of the carbon–energy tax in European Commission (1992d: Part C, 125).

18 EC (1992e: Article 9, 5–6). Member states were allowed to charge higher rates if they so chose.

19 EC (1992a: Article 10, 6). Whether EU exemptions from the tax under this article or Article 11 (see above) are consistent with GATT legal principles was not explicitly discussed by the Commission. See generally, Düerkop (1994).

20 Other factors contributing to emissions reduction were fiscal measures aimed at tax neutrality, including a fuel tax escalator, road charges and the elimination of the VAT exemption on domestic fuel and power (O'Riordan and Rowbotham 1996: 238–258).

21 Bundesministerium für Umwelt (1996: 8); for discussion of earlier voluntary programmes, including for returnable bottles, see Sydnes (1996: 211). See also the chapters in the other volumes of this series on the use and effectiveness of new regulatory instruments in the EU (Collier 1997, Golub 1998).

22 See German Ministry of the Environment *et al.* (1996: Part I, pp. 1–2 and Part 2, pp. 4–5).

23 See Sydnes (1996: 202 (Germany), 279 (Norway) and 304 (Italy)).

24 Most funding for the AIJ pilot phase has come from public agencies (MoH 1994: 71–72, German Ministry of the Environment *et al.* 1996: Part 2, pp. 21–23, Sydnes 1996: 279).

25 The Berlin Mandate in Decision 5/CP.1 established a two part regime for AIJ. Section (b) states that AIJ between Annex I and non-Annex I parties will not be seen as fulfilment of current commitments of Annex I parties under Article 4.2(b) of the Convention (ILM 1995: 1685). By negative implication trades between Annex I parties could fulfil current commitments, even in the absence of CoP accord on a JI protocol. Decision 5/CP.1 goes on to establish a pilot phase for AIJ between Annex I and non-Annex I parties, but emphasises in 1(f) that no credits shall accrue to any parties in pilot phase projects (ILM 1995: 1686). Since all EU member states, Eastern European and ex-Soviet nations in Europe including the Russian Federation are Annex I parties, intra-European trading would seem to occupy a favourable position under the FCCC. A reproduction of Annexes I and II to the FCCC may be found in Sydnes (1996: 380–381) or see ILM (1992: 872–873).

26 Tax and permit trading are economic instruments which may be distinguished on efficiency or equity grounds. Although they can yield identical results in theory, in practice uncertainty, administrative and institutional considerations may counsel the use of taxes over trading or vice versa (Weitzman 1974).

27 Taxes paid on infra-marginal emissions could be restored to emitters so that they bore only the cost of extra-marginal pollution as they would under grandfathering or other schemes of free distribution of the permit stock. Alternatively, if the stock of permits allowed was sold at public auction, the state could amass a fund equal to the full social cost of emissions collected under the carbon tax.

28 For discussion of constraints on climate change regime development specific to the multilateral negotiation forum, see Heller (1996: 330–331).

29 On the general problems of energy development markets in Asia, see, for example, Tam (1995). For explicit consideration of issues connected to energy plant development in China, see International Energy Agency (1996: 11–17). See also Heller (1995: 336).

30 Tellingly, although the Berlin Mandate denies credit for JI projects in meeting national commitments under Article 4.2, it allows them to be counted as technology transfer under Article 4.5. This concession has no significance for private actors. See ILM (1995: 1685).

31 For information on JI projects proposed and/or initiated under the pilot phase, see UNFCCC Secretariat (1996: FCCC/CP/1996/14 and Add. 1). See also Heller (1996: 298–299). For a contrasting view which warns against substituting a business for an aid perspective, see Chapter 8.

32 In theory, a two part test of policy and project baselines could be collapsed into one. To disqualify a JI project because the investment would have occurred in the absence of offset credits is to say that it has already been counted implicitly in the national (policy) baseline as part of the host jurisdiction's expected emissions trajectory. It should not be credited twice. At the same time, it should be stressed that organisational or institutional barriers to investment may impose risks that make seemingly attractive financial rates of return insufficient to attract actual investments by specific firms in particular markets. How these transactional risks are reflected in policy and project baselines will probably have to be settled at first on a case by case basis. Baselines, and thereby the standards for project qualification, may be elevated over time as experience in new markets and legal reforms weaken organisational and political barriers that earlier forestalled investment.

33 In a small numbers game like that envisioned among a committed core of Annex I participants, coordination costs ought not to be prohibitive. States may monitor each other's behaviour through informal mechanisms like the G-7 or by piggybacking on existing common institutions like the OECD.

Bibliography

Anderson, K. (1995) 'The Political Economy of Coal Subsidies in Europe,' *Fondazione Eni Enrico Mattei Working Paper Series*, Nota di Lavoro 40: 95.

Ausubel, J. and C. Marchetti (1996) 'Elektron: Electrical Systems in Retrospect and Prospect,' in special issue on 'The Liberation of the Environment,' *Daedalus*, Vol. 125, No. 3, pp. 139–169.

Barnett, S. (1992) 'Reaching a CO_2-emission Limitation Agreement for the Community: Implications in Equity and Cost-effectiveness,' *European Economy Special Edition*, No. 1, pp. 3–24

Bergesen, H., M. Grubb, J.-C. Hourcade, J. Jaeger, A. Lanza, R. Loske, L.A. Sverdrup and A. Tudini (1994) *Implementing the European CO_2 Commitment: a Joint Policy Proposal* (London: Royal Institute of International Affairs).

Beuermann, C. and J. Jäger (1996) 'Climate Change Politics in Germany,' in T. O'Riordan and J. Jäger (eds), *The Politics of Climate Change: a European Perspective* (London: Routledge).

Bruce, J., H. Lee and E. Haites (1996) *Climate Change 1995: Economic and Social Dimensions of Climate Change, Contribution of Working Group III to the Second Assessment Report of the Intergovernmental Panel on Climate Change* (Cambridge: Cambridge University Press).

Bundesministerium für Umwelt (1996) Bundesministerium für Umwelt, Naturschutz und Reaktorsicherheit, 'Annual Report for 1996 of the Federal Republic of Germany to the European Commission, Pursuant to the Council Decision of 24 June 1993 Concerning a System for Monitoring Emissions of CO_2 and Other Greenhouse Gases in the Community (93/389/EEC)'.

Climate Action Network and Climate Network Europe (1995) *Independent NGO Evaluations of National Plans for Climate Change Mitigation: OECD Countries, Third*

Review (1995) and Eop Group, Inc., Measuring up to the Year 2000 Aim of the Framework Convention on Climate Change.

Collier, U. (ed.) (1997) *Deregulation in the European Union: Environmental Perspectives* (London: Routledge).

Denis, C. and G. Koopman (1995) 'Differential Treatment of Sectors and Energy Products in the Design of a CO_2/Energy Tax for Employment, Economic Welfare and CO_2 Emissions,' *Fondazione Eni Enrico Mattei Working Papers Series*, Nota di Lavoro 32: 95.

Dinnage, J. and J. Murphy (1996) 'Treaty Establishing the European Community, Rome, 25 March, 1957, as amended by subsequent treaties,' *The Constitutional Law of the European Union: Documentary Supplement* (Ohio: Anderson Publishing Co.).

Dudek, D. and W. Wiley (1994) 'An Overview of Taxes and Trading as Environmental Control Policies,' in O. Hohmeyer and R. Ottinger (eds), *Social Costs of Energy: Present Status and Future Trends* (Berlin: Springer-Verlag).

Düerkop, M. (1994) 'Trade and Environment: International Trade Law Aspects of the Proposed EC Directive Introducing a Tax on Carbon Dioxide Emissions and Energy,' *Common Market Law Review*, Vol. 31, No. 2, pp. 807–844.

EC (1991) OJC L307/34.

—— (1992a) 'A Community Strategy to Limit Carbon Dioxide Emissions and Improve Energy Efficiency,' COM(92)246 final.

—— (1992b), 'The Economics of Limiting CO_2 Emissions,' Parts B and C, *European Economy*, Special Edition No. 1.

—— (1992c) 'The Climate Challenge: Economic Aspects of the Community's Strategy for limiting CO_2 Emissions,' *European Economy*, No. 51.

—— (1992d) 'Proposal for a Council Decision for a Monitoring Mechanism of Community CO_2 and other Greenhouse Gas Emissions,' COM(92)181 final.

—— (1992e) 'Proposal for a Council Directive Introducing a Tax on Carbon Dioxide Emissions and Energy,' COM(92)226 final, OJC C196/1.

—— (1993) *Towards Sustainability* (Brussels: Office for Official Publications of the European Communities).

—— (1995) 'Amended proposal for Council Directive Introducing a Tax on Carbon Dioxide Emissions and Energy,' COM(95)172 final.

—— (1996) 'Second Evaluation of National Programmes Under the Monitoring Mechanism of Community CO_2 and Other Greenhouse Gas Emissions,' COM(96)91 final.

Gerelli, E. (1991) 'Economic Responses to Global Warming: A European Perspective,' in R. Dornbusch and J. Poterba (eds), *Global Warming: Economic Policy Responses* (Cambridge: Cambridge University Press).

German Ministry of the Environment, German Ministry of Economics and German Industrial Association (1996) *Updated and Extended Declaration by German Industry and Trade on Global Warming Prevention*, issued March 27 at Cologne.

Golub, J. (ed.) (1997) *New Instruments for Environmental Protection in the EU* (London: Routledge).

Heller, T. (1995) 'Joint Implementation and the Path to a Climate Change Regime,' European University Institute, Robert Schuman Centre, Jean Monnet Chair Paper No. 23 (Florence: EUI).

—— (1996) 'Environmental Realpolitik: Joint Implementation and Climate Change,' *Indiana Journal of Global Legal Studies*, Vol. 3, No. 2, pp. 295–340.

Hoeller, P. and J. Coppel (1992) 'Energy Taxation and Price Distortions in Fossil Fuel Markets: Some Implications for Climate Change Policy,' *OECD Economics Department Working Papers*, No. 110 OECD/GD(92)70 (Paris: OECD).

Houghton, J. *et al.* (eds) (1996) *Climate Change 1995: the Science of Climate Change. Contribution of Working Group I to the Second Assessment Report of the Intergovernmental Panel on Climate Change* (Cambridge: Cambridge University Press).

ILM (1987) *International Legal Materials*, Vol. 26, No. 6, pp. 1541–1561.

—— (1991) 'Dispute Settlement Panel Report on United States Restrictions on Imports of Tuna,' *International Legal Materials*, Vol. 30, pp. 1594–1623.

—— (1992) 'United Nations Conference on Environment and Development Framework Convention on Climate Change, 9 May,' *International Legal Materials*, Vol. 31, No. 4, pp. 849–873.

—— (1995) 'United Nations Framework Convention on Climate Change Conference of the Parties, Decisions Adopted by the First Session, Berlin, 28 March–7 April,' *International Legal Materials*, Vol. 34, No. 4, pp. 1671–1710.

International Energy Agency (1995a) *Industry Attitudes to Steam Cycle Clean Coal Technologies: Survey of Current Status* (Paris: OECD/IEA).

—— (1995b) *Factors Affecting the Take-up of Clean Coal Technologies: Overview Report* (Paris: OECD/IEA).

—— (1996) 'Activities Implemented Jointly Under the FCCC,' *Extract from the Energy and Environment Update,* No. 3.

Koutstaal, P. and A. Nentjes (1995) 'Tradable Carbon Permits in Europe: Feasibility and Comparison with Taxes,' *Journal of Common Market Studies*, Vol. 33, No. 2, pp. 219–233.

Liberatore, A. (1995) 'Arguments, Assumptions and the Choice of Policy Instruments: The Case of the Debate on the CO_2/Energy Tax in the European Community,' in B. Dente (ed.), *Environmental Policy in Search of New Instruments* (Dordrecht: Kluwer Academic Publishers).

Marchetti, A. (1996) 'Climate Change Politics in Italy,' in T. O'Riordan and J. Jäger (eds), *The Politics of Climate Change: a European Perspective* (London: Routledge).

MoH (1994) 'Netherlands' National Communication on Climate Change Policies,' Ministry of Housing, Climate Change Division, GX, The Hague.

Nakicenovic, N. (1996) 'Freeing Energy from Carbon,' in special issue on 'The Liberation of the Environment,' *Daedalus*, Vol. 125, No. 3, pp. 95–112.

O'Riordan, T. and E. Rowbotham (1996) 'Struggling for Credibility: The United Kingdom's Response,' in T. O'Riordan and J. Jäger (eds), *The Politics of Climate Change: A European Perspective* (London: Routledge).

Richels, R. *et al.* (1996) 'The Berlin Mandate: The Design of Cost-Effective Mitigation Strategies' (paper on file with author).

Skjaerseth, J. (1994) 'Climate Policy of the EC: Too Hot to Handle,' *Journal of Common Market Studies*, Vol. 32, No. 1, pp. 25–45.

Smith, C. (1995) 'Carbon Leakage: A Review and Empirical Assessment,' *Fondazione Eni Enrico Mattei Working Paper Series*, Nota di Lavoro 10: 95.

Sydnes, A. K. (1996) 'Norwegian Climate Policy,' in T. O'Riordan and J. Jäger (eds), *The Politics of Climate Change: A European Perspective* (London: Routledge).

Tam, C. (1995) 'Features of Power Industries in Southeast Asia: Study of Build-Operate-Transfer Power Projects in China,' *International Journal of Power Management*, Vol. 13, pp. 303–311.

Tietenberg, T. (1985) *Emissions Trading: An Exercise in Reforming Pollution Policy* (Washington, DC: Resources for the Future).

Tripp, J. and D. Dudek (1989) 'Institutional Guidelines for Designing Successful Transferable Rights Programs,' *Yale Journal of Regulation*, Vol. 6, No. 2, pp. 369–91.

UNFCCC Secretariat (1996) 'Activities Implemented Jointly: Annual Review of Progress under the Pilot Phase and Addendum,' FCCC/CP/1996/14 and Add. 1, 4 June.

Watson, R., M. Zinyowera and R. Moss (1996) *Climate Change 1995: Impacts, Adaptations and Mitigation of Climate Change, Contribution of Working Group II to the Second Assessment Report of the Intergovernmental Panel on Climate Change* (Cambridge: Cambridge University Press).

Weitzman, M. (1974) 'Prices vs Quantities,' *Review of Economic Studies*, Vol. 41, No. 4, pp. 477–491.

Welsch, H. (1995a) 'Joint vs Unilateral Taxation in a Two Region General Equilibrium Model for the European Community,' *Fondazione Eni Enrico Mattei Working Paper Series*, Nota di Lavoro 36: 95.

—— (1995b) 'The Carbon Tax Game: Differential Recycling in a Two-Region General Equilibrium Model of the European Community,' *Fondazione Eni Enrico Mattei Working Paper Series*, Nota di Lavoro 37: 95.

World Bank (1994), 'China: Issues and Options in Greenhouse Gas Emissions Control' (Washington, DC: World Bank).

Yun, L. (1996) 'Opportunities and Concerns in the Developing World,' Paper presented at Chatham House, London, 5 December.

6

EU ENVIRONMENTAL POLICY AND THE GATT/WTO

David Vogel

This chapter explores the relationship between the regulatory policies of the European Union and the Union's trade relations with the United States and other countries. The first section examines the general issue of trade and environment, placing it within the context of the rules and procedures of the General Agreement on Tariffs and Trade (GATT)/World Trade Organisation (WTO). The second section discusses two of the most visible trade disputes over the alleged use of environmental regulations as trade barriers that have involved the European Union. Both disputes involved the United States: the first stemmed from the impact of American automotive fuel economy standards on European luxury car exports to the US and the second from an American ban on imports of European processed tuna fish. The chapter then turns to a series of ongoing trade conflicts in which the EU has been accused of engaging in unfair trade practices to promote conservation, protect wildlife and promote ecolabelling. In light of these disputes, the concluding section assesses the EU's recent proposals to reform the environmental provisions of the GATT/WTO and draws some general conclusions about the development of EU policies and practices on trade and environmental protection.

The growing linkages between trade policies and environmental regulations in recent years stem from the convergence of two policy objectives: trade liberalisation and environmental protection (Vogel 1995). The increase in economic integration has subjected a growing number of national policies, including environmental regulation, to greater international scrutiny. These regulations are rarely neutral and often, either intentionally or unintentionally, disadvantage foreign producers. At the same time, the number of environmental regulations has continued to expand. Many of these regulations, especially those directed at protecting the global commons, include trade restrictions, either because the harm they address is trade-related or because they are enforced through trade sanctions. Consequently, there is frequently a tension between reducing trade barriers and strengthening environmental standards.

Since there is often substantial disagreement as to whether or not a particular environmental regulation constitutes an unreasonable interference with trade, balancing trade liberalisation and environmental protection poses a difficult challenge. As the world's foremost effort to promote trade liberalisation, the EU has been wrestling with this challenge for at least two decades. The European Court of Justice has developed an extensive body of jurisprudence which defines the circumstances under which the environmental regulation of a member state violates its treaty obligations, while the Council of Ministers has established a number of uniform regulatory standards in order both to maintain the single market and strengthen European environmental quality.

By contrast, the GATT/WTO has experienced much more difficulty in balancing these two policy objectives. Unlike the EU, the GATT/WTO lacks the authority to impose environmental regulations. As a more specialised and much weaker organisation, the GATT/WTO only has the authority to determine when a national, or in the case of the EU, regional environmental standard, violates the principles of free trade. Moreover, while a commitment to strengthening environmental regulations was incorporated into the provisions of the Single European Act of 1986, the first mention of the word 'environment' in the GATT/WTO did not occur until the Uruguay Round agreement of 1994.

The preamble to the Standards Code, which was incorporated into the provisions of the newly established WTO, states that each country 'may maintain standards and technical regulations for the protection of human, animal, and plant life and health and of the environment' (Vogel 1995: 136). The agreement also requires that national standards or 'technical barriers to trade' 'not be more trade-restrictive than necessary to fulfil a legitimate objective, taking into account the risks non-fulfilment would create' (Vogel 1995: 136). As we shall see, these two provisions leave a number of critical questions unanswered.

Trade disputes

Fuel economy

The most important environmentally-related trade dispute involving the EU occurred in 1993, when the EU requested the convening of a dispute settlement panel to rule on the GATT consistency of American corporate average fuel economy (CAFE) standards, the so-called gas-guzzler tax, and a tax on luxury automobiles. The EU's complaint centred on the fact that all three burdens fell disproportionately on European companies. While European cars accounted for only 4 per cent of American sales in 1991, they contributed 88 per cent of the revenues collected by American luxury taxes and CAFE penalties, totalling $494 million (Lavelle 1993: 39). The purpose of the CAFE standards, originally established in 1975 and subsequently tightened in 1980, is to promote fuel efficiency. They are based on the miles per gallon achieved by a sales-weighted average of all vehicles produced by a manufacturer. If a manufacturer's vehicles fall below this standard (which in 1993 was 27.6 miles per gallon) they face a penalty of $5 for every tenth of a mile per gallon, multiplied by the number of automobiles sold in the United States.

Although this penalty applies equally to all car manufacturers, it has been paid exclusively by European limited-line premium car makers. Since all American automobile firms make a full line of cars, they have avoided the tax on their less fuel-efficient luxury cars by averaging their fuel economy with the rest of their fleet. Japanese car firms have avoided the tax because they mostly make smaller, more fuel-efficient vehicles. According to A. B. Shuman, public relations manager for Mercedes-Benz of North America, CAFE rules 'are really made for the Big Three. The problem for Europeans is that . . . they don't have little cars to balance out the higher-consumption cars' (Lavelle 1993: 39).

In its complaint to the GATT, the EC argued that 'the CAFE regulations are biased toward the full-line manufacturers and limited-line manufacturers that produce mostly small vehicles' (Dunne 1993: 5). The EC claimed that because CAFE penalties fell only on imported cars, they violated the GATT's national treatment provision. The EC further argued that the 1978 'gas-guzzler tax,' which was based on a threshold fuel economy standard of 22.5 mpg, was not only discriminatory but was not based on any objective or reasonable criteria. Finally, the EU claimed that the $30,000 cutoff for the luxury tax was both 'capricious' and discriminatory, since in 1990, the year the tax was introduced as part of the 1990 budget reconciliation bill, more than 80 per cent of the vehicles subject to it were imports (TE 1993: 4).

In the Autumn of 1994, the GATT dispute panel found all three taxes to be GATT consistent. The panel noted that while the EU was correct in suggesting that the American policy objective of promoting fuel efficiency could be achieved in ways that were less trade-restrictive, namely by increasing gasoline taxes, it declined to hold the United States to a 'least trade restrictive' standard. Rather it concluded that all the United States needed to demonstrate was that its regulation achieved a legitimate environmental objective, a category in which fuel conservation fell. The panel also upheld the GATT consistency of both the gas-guzzler and luxury taxes, since they were levied on products rather than on companies.

The United States was extremely pleased with the GATT panel ruling. According to USTR head Micky Kantor,

The panel has emphatically rejected the Europeans' claim that trade-neutral legislation intended to further energy conservation goals and protect the environment could be attacked because

Chrysler, Ford and GM invested and complied with the law while Mercedes and BMW chose not to and had to pay penalties.

(IUST 1994: S-1)

However, the European Commission characterised the panel report as a 'backward step in the interpretation of GATT Article III that risks opening the door for inventive tax and regulatory authorities to discriminate against imported products' (Charnovitz 1996: 461).

The dispute panel did rule that the provision of the CAFE legislation requiring companies to meet CAFE averages for both their domestic and imported cars (the so-called separate fleet accounting rules) was GATT inconsistent, since it treated similar products differently on the basis of where they were manufactured. The United States refused to change this provision on the grounds that it had no adverse impact on European companies, since all their cars were imported. Nonetheless, the EU criticised the US action. As one EU official put it: 'We object to the fact that the US appears unwilling to change the CAFE law. . . . Such a posture by the US does not bode well for the future of the multilateral trade system of the World Trade Organisation' (IUST 1994: S-7).

Tuna–dolphin

One of the most contentious sources of conflict between GATT/WTO rules and environmental regulations has involved the use of import restrictions to promote global conservation. During the 1970s, in order to protect dolphins in the eastern tropical Pacific from being killed by tuna fishermen, the United States established a dolphin kill quota for its fishing fleets. In 1990, as a result of a federal court decision, this regulatory standard was applied to foreign tuna vessels as well. Consequently, the United States banned imports of tuna from Mexico and Venezuela on the grounds that the fishing practices of their fleets violated American standards for dolphin protection. At the same time, the United States also prohibited tuna imports from Costa Rica, France, Italy, Japan and Panama since these countries purchased tuna from the nations subject to a direct embargo.

Mexico immediately filed a complaint with the GATT. It claimed that GATT rules prohibit a nation from restricting the import of a product on the basis of how it was produced outside its borders. Mexico's complaint was supported by the European Union, which reiterated its long-standing objections to American unilateralism in pursuit of global environmental objectives (Maggs 1992: 3A).

In April 1991, a GATT dispute panel ruled that the American trade embargo violated its GATT obligations, since, according to the trade agreement's 'national treatment' provision (Article III), imports can only be restricted on the basis of characteristics of the product itself – provided similar restrictions are imposed on domestically-produced ones. The GATT does not permit signatory nations to 'restrict imports of a product merely because [the product] originates from a country with environmental policies different from its own' (Housman and Zaelke 1992: 10274).

Following the passage of the International Dolphin Conservation Act of 1992, the United States informed France, the United Kingdom and several other countries that they had been removed from the list of nations subject to the intermediary embargo as they had certified that they were no longer importing yellowfin tuna from the nations still subject to the direct embargo. However, the American secondary embargo continued to be applied to tuna imports from four nations: Spain, Italy, Costa Rica and Japan.

The EU had criticised Mexico's refusal to submit the tuna–dolphin dispute panel report to the GATT Council, which would have enabled it to become officially adopted. It argued that 'what had started as a dispute between two parties was now of interest to us all', and demanded that the GATT Council hold a full debate on the tuna report in order to correct the false 'impression in some quarters that the report had placed environmental and trade issues on a collision course' (GF 1992: 5). Frustrated by the fact that the panel report enjoyed no legal standing, the European Union, acting on behalf of the two member states still affected by the American embargo, requested a second dispute resolution panel in June 1992.

In its brief the EC stated that 'while it agrees with the environment goal being pursued by the US, it objects to the US imposing its laws on the rest of the world' (GF 1992: 5).[1] It also claimed that it had suffered substantial economic injury, because tuna fish that otherwise would have been sold in the United States was flooding the world market, thus lowering tuna prices outside the United States.

In its submission to the panel, the United States argued that its secondary embargo fell well within the terms of Article XX, since it was intended to protect an 'exhaustible natural resource', whose physical presence on the high seas made 'them more rather than less in need of conservation' (GF 1992: 5). Furthermore, it had been imposed in conjunction with restrictions on domestic production and consumption; indeed, far from being protectionist, the United States had imposed even more stringent requirements on its own fishing fleets. The United States further argued that 'there is nothing in the General Agreement that distinguishes between "unilateral" measures and other types of measures'. In fact, the term 'unilateral' appears nowhere in the General Agreement. Moreover, 'the vast majority of measures taken by sovereign nations in all fields of activity are unilateral' (Marshall 1992: A8). According to the Americans, 'at stake is the ability of a nation to take measures to protect global resources' (Fraser 1992: 4).

In June 1994, a GATT dispute panel also found the secondary embargo to be GATT inconsistent. As in the case of the first tuna–dolphin dispute, the United States refused to comply with the panel ruling. However, in

1995, the United States, along with Costa Rica, France, Mexico, Spain and Venezuela, signed the Panama Declaration. This is an international agreement which limits the total number of dolphin deaths as well as protecting sea turtles and small fish (ENS 1995). Once this agreement is ratified by the US, American restrictions on tuna imports will be lifted.

Ongoing trade disputes

Animal protection

Notwithstanding its harsh criticism of American unilateralism, the EU has frequently engaged in similar practices. In 1983, following a number of resolutions by the European Parliament that drew public attention to the hunting methods used to cull baby seals off the Canadian coast, the European Council banned the import of skins of pup harp and hooded seals, as well as products derived from them (Freestone 1991: 141, Demaret 1993: 328–329). The Council, hesitant to directly threaten Canada's economic interests, justified its restriction not on humanitarian grounds but on the need to preserve the common market, since a number of member states had enacted similar bans.

The Canadian Government strongly protested the EU's action and filed a complaint with the GATT on the grounds that the General Agreement only permits a nation to ban unilaterally the import of a product which is only produced outside its borders if the product's consumption harms its own citizens or environment. Canada subsequently withdrew its complaint and shortly thereafter announced an end, on conservation grounds, to all commercial hunting of seal pups. When the EU extended the ban in 1988, the European Commission noted that harp seal pup hunting in both Canada and Norway had significantly declined as a result both of its 1983 Directive and the hunting restrictions the two countries had adopted.

In 1991, the European Council again approved a trade restriction to protect wild animals outside its legal jurisdiction (Demaret 1993: 330). The Council banned leghold traps to catch wild animals within the Union and, beginning in 1995, prohibited the importation of thirteen species of fur from nations that continued to use leghold traps, rather than more 'humane trapping standards' (Simon 1991: 2). Like the restriction on imports of products made from baby seals, this policy was motivated by an intense campaign on the part of European animal rights groups, whose supporters including the actress turned activist Brigitte Bardot. However, the ban was also supported by Europe's own large pelt industry, which relies upon 'farming' rather than trapping. In part due to anti-fur movements in a number of European countries, this industry has experienced a severe decline in sales.

The ban was primarily aimed at the United States, Canada and the countries belonging to the former Soviet Union – the main suppliers of pelts of wild mammals to the Union's fur industry. The EC's decision provoked a particular outcry from Canada, where approximately 80,000 trappers earn more than $30 million a year from pelt exports to Europe (Simon 1991). Half of these trappers are aborigines, many of whom depend on trapping for most of their income; a delegation of native Canadians accused the EU of committing 'cultural genocide' (Southey 1995: 4). Approximately 70 per cent of the furs harvested in North America are exported to Europe.

Caught between pressures from the European Parliament and animal welfare groups on one hand, and anxious to avoid a trade dispute with the United States and Canada on the other, in November 1995, the European Commission decided to postpone the import ban for a year to allow its trading partners to find a more humane alternative to leghold traps and reach agreement on an international standard for fur harvesting. The Commission also exempted from the ban furs from animals caught by indigenous people. The Commission's announcement was immediately denounced by European animal welfare activists (EE 1996). On 14 December 1995, the European Parliament condemned the Commission's action by a vote of 262 to 46; it called upon the EC to institute a ban immediately. In language reminiscent of the response of American environmentalists to the GATT tuna–dolphin decision, the Commission was accused of 'sacrificing animals on the altar of free world trade' (AE 1996).

In March 1996, EU environment ministers, frustrated by the lack of progress on an international standard, warned exporters that the Union would implement its ban in December unless discussions aimed at finding an alternative to leghold traps produced satisfactory results (FT 1996: 5). They also announced their opposition to exempting furs and skins caught by indigenous populations from a ban. One member state, the Netherlands, chose to ignore the Commission and instituted an import ban on furs and skins from animals caught with leghold traps. The United States and Canada responded to the Dutch action by threatening to file a complaint with the WTO (UPI 1996). In May 1996, a delegation of fur trappers from North America arrived in Brussels to lobby the EP against the ban, claiming that they have developed new types of traps that are less cruel to animals. For its part, the Commission is continuing to work with the International Standards Organisation to develop a definition of 'inhumane traps', since according to the WTO Standards Code, a trade restriction based on a standard which is internationally recognised is more likely to be found WTO consistent.

Forest protection

The European Parliament has also called upon the Union to halt all imports of tropical hardwoods originating in Sarawak, Malaysia, on the grounds that logging in that region is proceeding at 'a devastating pace' (Demaret 1993: 334–335). It has also requested that the Union promote conservation by limiting the importation of hardwoods to countries which had adopted 'Forestry Management and Conservation Plans'. The effect of the latter proposal would be to involve the Union in negotiating quotas with each producing country, which it already does for a number of agricultural products, though for economic rather than environmental reasons.

However, both the Council and the Commission have resisted these pressures from the EP on the grounds that it would be inappropriate for the Union to act outside of the international frameworks provided by the International Tropical Timber Agreement, the GATT, or CITES. According to the Commission, 'Import restrictions unilaterally applied by the Community might . . . benefit other consuming countries without doing much to save tropical forests from destruction' (Demaret 1993: 334–335). But the EU's foreign trade ministers have called for a system of timber identification aimed at protecting tropical forests from over-exploitation and destruction under the auspices of the International Timber Organisation. Their proposal was denounced by Malaysia, which accused the EU of seeking to 'bash' tropical timber exporting countries in order to protect its domestic forestry industry (Pleydell 1993).

Consistent with this policy, the EU strongly criticised a 1985 Indonesian ban on the export of unprocessed tropical timber from its rainforests. Disputing Indonesia's claim that the 'ban [was] an integral part of a long-term programme for conservation and management of its forest resources', the EU claimed that its real purpose was to protect Indonesia's wood processing industry, by requiring that the manufacture of 'value added' products such as furniture take place in Indonesia rather than in Europe or Japan (Hegenbart 1991: 231). They argued that the export ban violated Article I of the GATT, which requires foreign and domestic firms to be treated equally.

Ecolabelling

Ecolabelling has emerged as an increasingly important instrument of environmental policy, especially in Europe, where significant numbers of consumers have indicated their interest in purchasing 'green' products. A 1995 survey reported that 82 per cent of German consumers, 67 per cent of Dutch consumers and 50 per cent of French and British consumers stated that they 'incorporate[d] environmental concerns in their shopping behaviour' (Rosen and Sloane 1995: 76). According to another study, '67 per cent of EU citizens had already purchased or were ready to buy "green" products' (PRN 1996a: 1). Ecolabelling thus can offer European firms an important market incentive to improve their environmental practices. They can strengthen their environmental practices while maintaining their market shares.

The first and most successful national ecolabelling scheme, known as the 'Blue Angel', was initiated by Germany in 1977. Subsequently the Nordic Council established a 'White Swan' programme and a number of other EU member states, including Great Britain, France, the Netherlands and Spain organised their own ecolabelling schemes (see Eiderstrom 1998). All are voluntary and are administered through non-state bodies.

The EU has become concerned about the proliferation of national eco-labelling programmes because fifteen different national labels, each using distinctive criteria and designating different products, would confuse consumers. Furthermore, if national labelling bodies employed criteria that were more difficult for producers in other member states to meet, eco-labelling could undermine the single European market. In 1992, to address these problems, the EU authorised a voluntary 'ecolabel' to be awarded to ecologically sensitive products that met objective environmental standards established at the EU level (Bristow 1994: 50–55, Eiderstrom 1998).

This ecolabel was not intended to replace national labels, but rather to supplement them, though the Commission assumed that many manufac-turers would prefer the ecolabel because it would be recognised throughout the Union. Manufacturers were first required to apply to a national body in either the member state where the product was manufactured or, in the case of imports, where it was first imported. If their product meets the criteria, an ecolabel would be awarded which the manufacturer or distributor could then use throughout the Union. The Commission was instructed to apply a 'cradle to grave' assessment, meaning that the ecological impact of each product would be evaluated throughout its production cycle – from the extraction of raw materials to its disposal after use.

The standards for awarding EU ecolabels are established jointly by the EU and the member states (NWF 1996: 13–14). After the EU decides which specific product categories are to be included in the programme, it assigns each to a member state in order to conduct life-cycle assessment research. The 'lead country' then develops product criteria based on this research. However, these product criteria must then be approved by all member states in order to prevent any distortion of the single market. The member states are also responsible for processing applications for certifica-tion, awarding labels and monitoring label use within their borders.

However, the establishment of 'euro-labels' has proceeded slowly. By November 1994, standards had only been established for washing machines, dishwashers, kitchen rolls, toilet paper and soil improvers, while to date only a handful of products with European 'ecolabels' are actually on sale in Europe (Harding 1995: 10). Some national representatives have suggested reducing the number of ecolabelling criteria for each product group in order to expedite the approval process (EW 1996a: 11–12).

One of the difficulties the Union has faced stems from differences in priori-ties and practices among the fifteen member states. Packaging standards are a case in point.[2] The EU's northern member states, namely Germany, the Netherlands and Sweden, argue that the extensive use of packaging produces large amounts of solid waste and wastes resources. Accordingly, they have insisted that the EU's 'eco-daisy' label, which is meant to encourage the 'environmentally-friendly' packaging of all products sold within the EU, be granted only to manufacturers who use both recycled and recyclable material in their packaging cartons or containers. However, the use of this criterion was strongly opposed by the Union's poorer, southern member states who contend that traditional packaging helps sell their products and is more cost-effective than recycling.

The EU's ecolabelling programme has also created tension between the Union and its trading partners (Jha and Zarrilli 1994: 64–73). A number of the latter have expressed concern that both national and Union green labelling programmes might serve as trade barriers. For example, wood pulp and paper producers from the US, Canada and Brazil claim that the EU's criteria for 'green' kitchen rolls and toilet paper constitute non-tariff trade barriers, as they place too much emphasis on the use of recycled materials and not enough on appropriate forest management. According to an official from the American Forest and Paper Association,

> Just because a paper is recycled does not mean that it has less of an impact on the environment. We have programmes on sustainable forestry that we feel are just as safe for the environment. . . . We feel that the Commission's criteria is in fact designed to subsidise the recycling industry in Europe.
>
> (BNA 1996a: 935)

If a WTO complaint is filed by one of the EU's trading partners, it may well revolve around the Union's paper standards (BNA 1996a).

An article in *Newsweek* in the spring of 1996 entitled 'Seeing Red Over Green' highlighted the growing fears of much of the American business community over the EU's ecolabelling programme. It noted:

> Paper recycling might make sense in Holland, but requiring paper made in Canada's sparsely settled west to use recycled pulp may consume more resources than it saves. Or take the EU's eco-label for T shirts. US makers claim the rules permit more pollution from plants that dump wastewater into the sewer than do those that treat it on site, as most US textile plants do.
>
> (Levinson 1996: 55)

The EU's ecolabelling criteria have come under especially strong criticism from a number of Third World countries, who fear that the labels would be employed as a form of 'green protectionism' (see also Chapter 8). For example, Brazil's wood producers have criticised the EU's unwillingness to consider paper produced from damaged wood or sawdust as having been made from 'renewable resources'. ABECEL, the association of Brazilian wood exporters, has urged the EU to revise its criteria for paper products in order to achieve a better balance between the promotion of recycling and sustainable forest management. ABECEL informed the EU that 'establishing recycling as the overwhelming dominant criterion for judging the eco-label, to the exclusion of other principles, is more likely to damage the cause of environmental protection than to help it' (Banki 1995: 7). Likewise, Third World producers have criticised the EU's efforts to establish a life-cycle standard for textiles that includes restrictions on the use of pesticides in cotton production. They claim this production standard will be used to protect European firms from the phasing out of the Multi-Fiber Agreement which has limited Third World textile exports to Europe.

From the perspective of European producers, it makes sense to use the ecolabel to reward European firms who have been required to devote considerable resources to address European environmental problems, such as acid rain. By this logic, to label an imported product 'green' if it were produced in ways that resulted in substantial emissions of sulphur – even if acid rain was not a problem in the exporting country – would unfairly place European producers at a competitive disadvantage. Not surprisingly, this logic has been sharply criticised by many of the EU's trading partners, who argue that they should not be required to adopt the EU's production standards to earn a 'green' label. Similar conflicts have emerged over the application of EU production standards to paper and leather products produced outside the EU.

The trade implications of the EU's ecolabelling plans have come before the WTO. In response to widespread criticism of the GATT's decision in the tuna–dolphin case, in 1991 the GATT agreed to reconvene its Working Group on Environmental Measures and International Trade. This group was reconstituted as the Committee on Trade and Environment (CTE) following the establishment of the World Trade Organisation in 1995. Among the most important issues facing the Committee has been the WTO consistency of ecolabelling standards. Specifically, the CTE has attempted to reconcile the growth of ecolabels with the newly strengthened Standards Code that commits all WTO signatories to ensure that 'technical regulations and standards, particularly with respect to the packaging ... and labelling of goods ... create no unnecessary obstacles to international trade' (Banki 1995: 15). Among the issues recently discussed by the WTO's Committee on Trade and Environment are mutual recognition, certification criteria and transparency issues for ecolabelling.

There is considerable debate about whether the European ecolabel falls within the scope of the Standards Code (Banki 1995). The European ecolabel is clearly not a 'technical regulation' because compliance is voluntary. However, because the EU is a widely recognised organisation and governments play an important if informal role in determining the criteria for awarding ecolabels, its label could constitute a 'standard', which would bring it within the code's scope. This would mean that the EU would be required to make its labels non-discriminatory, transparent, and based whenever possible on international standards. It would also be obligated to provide its trading partners with the opportunity to comment on the criteria used to award ecolabels.

The WTO wishes to encourage the use of ecolabels since they are clearly preferable to unilateral barriers to trade, such as the American tuna embargo. They also represent a promising new approach to environmental regulation, one which relies on market incentives rather than command and control. Significantly, in the tuna–dolphin case, the GATT dispute panel found the American 'Dolphin-safe' labelling requirement to be GATT consistent, even though it was based on a process, rather than a product standard, and sought to influence environmental conditions outside the legal jurisdiction of the United States. The panel reasoned that the American labelling standard was GATT consistent because it applied to all tuna fish caught in a given area and did not distinguish the products on the basis of national origin. Also, unlike the embargo, it did not prevent consumers from purchasing tuna regardless of how it was caught.

Still, the WTO, as well as the Organisation for Economic Cooperation and Development, fears that the growing use of ecolabels by developed countries may restrict trade. In particular, life-cycle based labelling programmes, which incorporate criteria relating to process and production methods, frequently appear to favour environmental conditions and preferences in the importing country. A central holding in the 1991 GATT dispute panel decision in the tuna–dolphin dispute between the United States and Mexico was that the GATT did not permit nations to restrict imports on the basis of how a product was produced or processed outside its own borders. From this perspective, the EU's use of life-cycle criteria for the awarding of ecolabels represents a 'slippery slope' which could lead to legitimating the use of process production measures (PPMs) as trade barriers.

For its part, the US has claimed that the EU's ecolabelling programme 'violates the law of international trade', and it has threatened to file a formal complaint with the WTO (Kirwin 1996: 447). According to the US ambassador to the EU, Stuart Eizenstat, the American concern was 'not with eco-labels per se. Our problem is that the process is not sufficiently transparent and does not allow for the participation of non-EU industries' (Kirwin 1996). He added that he was also concerned about the problem of discrimination:

> We want to be able to show that if we have equivalent environmentally benign production processes, they should be given an eco-label. Just because a US process is different does not mean that

it is environmentally unsound. Our hope is that this does not turn into a covert trade restriction by favouring EU processes.

(Kirwin 1996)

An American trade association official has stated that 'this is a very important issue not only for the US but for countries such as Canada and Brazil' (BNA 1996b: 884). An American trade lawyer has characterised both the EU's packaging and labelling standards as an 'area of enormous potential conflict', adding that 'as we move toward international standardisation of labelling, Europe is going to be setting the agenda' (BNA 1995). Nonetheless, the United States has yet to take any formal action under the WTO, largely because the European labelling programmes have yet to affect adversely the sales of any American product in Europe.

On 28 February 1996, the EU Commission issued a statement encouraging the CTE to 'work on a WTO regime that ensures full transparency and non-discrimination in voluntary eco-labelling schemes' (PR 1996). The EU has specifically proposed that the Technical Agreement to Trade agreement be amended to permit the use of ecolabelling schemes based on a life-cycle approach, thus assuring the WTO consistency of its own scheme. This proposal has meet with opposition from many Third World nations since it would legitimate the application of process and production requirements, albeit voluntary ones, to traded products. However, such an amendment would also subject ecolabelling to WTO discipline.

At the same time, a subcommittee of a working group of the International Standards Organisation (ISO), a private standards setting body based in Geneva, has begun work on international standards for ecolabels. These would emphasise 'credibility, consultation with stakeholders, transparency, accessibility and avoiding the creation of unnecessary obstacles to trade'.[3] The working group's efforts are strongly supported by the United States and export-reliant developing countries who fear the use of ecolabels as trade barriers. However, the development of an international consensus on an ISO Standard on Environmental Labels and Declaration (Draft Standard 14020) has proven difficult. In particular, a number of national delegations, including those from the UK, Germany and France, want the ISO's draft principles to serve as guidelines rather than standards, a position which the United States and Canada strongly oppose (EW 1996b: 1–2).

There are, however, other ways of promoting international cooperation on ecolabelling. These include, for example, employing the principle of 'equivalency' which would permit a product to qualify for an ecolabel if it meets minimal or equivalent environmental standards, even if they were not identical to those of the country to which the product was exported. Another approach is 'mutual recognition', which would mean that the importing country's ecolabel could be awarded to any products covered by the ecolabel of the exporting country. All of these approaches remain at the discussion

154

stage.[4] The United Nations has begun work on devising a certification scheme which would reflect the environmental and economic priorities of both developed and developing countries (EW 1994).

Trade and environment

In March 1996, the European Commission issued a report on trade and environment to the EU Council as part of its preparation for the first ministerial meeting of the WTO in Singapore in December 1996. It began by asserting that there was in principle no tension between a liberal trading system and environment regulation: not only had a $250 billion market developed in green technology, but it estimated the EU's costs of compliance with strict environmental standards at only between 1 and 2 per cent of overall production costs. In short, the EC claimed that strict regulatory standards were *not* a source of competitive disadvantage. However, the EC cautioned that 'friction can arise when domestic rules hamper or discriminate against imports . . . [or when] nations tackle transboundary problems by setting rules beyond their jurisdiction' (PRN 1996b: 1). The former includes such measures as 'life-cycle' ecolabelling and recycling requirements. The EC strongly endorsed the use 'life-cycle' ecolabelling requirements while at the same time emphasising the need for international rules to assure they are implemented in ways that are both transparent and non-discriminatory.

In the case of extra-jurisdictional environmental measures, the Commission urged the WTO to recognise the legitimacy of trade measures taken under multilateral environmental agreements and urged the EU to encourage environmental improvements in developing nations through market premiums and preferential access rather than eco-duties. Nonetheless, the EC did acknowledge that there might be 'exceptional cases' that would justify trade restrictions taken outside a multilateral framework – for example, if a state had 'breached its obligation toward fundamental international environmental law and the health of the world ecosystem' (PRN 1996b: 2).

The Commission's statement about the essential compatibility of free trade and environmental regulation clearly represented an effort to stake out the environmental 'high ground'. Its efforts to make sure that WTO rules and procedures do not interfere with the strengthening of national, regional or global environmental standards, popularly referred to as 'greening the GATT', reflects the continuing strength of green parties and pressure groups in much of Europe as well as their substantial influence in the European Parliament. It also demonstrates the extent to which global environmental leadership has passed from the United States to the European Union.

While American environmental officials have been preoccupied in recent years with preventing the rolling back of existing environmental rules and regulations, EU environmental policymaking has been relatively innovative, developing initiatives such as ecolabelling and integrated pollution control. The EU has also taken the initiative in developing 'greener' international standards, such as for ecolabelling and fur trapping.

The leadership role played by the Commission in Geneva contrasts with that of the United States, whose role in the deliberations of the WTO's Trade and Environment Committee has been much more passive. One WTO official observed, 'The US is proposing nothing and systematically trashing everyone else's proposals. It is a major obstacle to getting anything done' (Williams 1996: 3). Ironically, while it was American pressure that led to the establishment of the trade and environment committee in the first place, the United States has yet to advance its own policy proposals on any of the critical issues on the committee's agenda. Caught between the Republican congressional victory in the 1994 mid-term elections and the impending presidential election, and fearful of antagonising either environmentalists or industry, as of late 1996 the Clinton Administration had yet to propose any changes in WTO rules or dispute procedures that would strengthen the links between trade liberalisation and environmental improvement.

The Commission's statement also reflects two of the EU's most important priorities. Most importantly, it wants to make sure that its life-cycle eco-labelling programme is not challenged by the WTO. This programme is not only important to European environmentalists but also to European industry, which is faced with the significant costs of complying with EU environmental directives. Ecolabelling represents a strategy for enabling European consumers to share in these costs by promoting the 'greener' products produced by European firms. It is a critical component of the EU's efforts to improve European environmental quality without unduly impairing the competitiveness of European industry.

Second, by supporting the establishment of new procedures for handling of disputes that might arise out of the trade provisions of multilateral environmental agreements, the EU accomplishes two objectives: it reinforces its longstanding opposition to American efforts to use trade restrictions to enforce extra-jurisdictional environmental regulations in the absence of international environmental agreements, and it protects European firms from imports from countries which have not complied with international environmental agreements to which the EU is a signatory. The latter issue is likely to become especially critical if and when an international agreement to curb carbon emissions is adopted.

The EU's efforts to reconcile the political demands from European environmentalists with the welfare of European firms is apparent in its position on each of the trade and environment conflicts in which it has been involved. The initial pressures for the EU's initiatives regarding imports of hardwoods from tropical forests, furs from animals caught using leghold traps and ecolabelling all came primarily from environmentalists. At the same time, it is noteworthy that none of these policies disadvantages

European producers. On the contrary, all would either maintain or increase their European market shares. Thus both the Union's standards for paper products and its efforts to restrict the imports of hardwoods would benefit European wood processing firms and its forestry industry. Likewise, any restrictions imposed on fur imports would assist Europe's financially hard pressed fur industry. Most importantly, both European and national eco-labelling criteria, no matter how transparent or 'non-discriminatory', are far more likely to promote the sales of European products than of imported ones.

At first glance, it is difficult to find a logically consistent pattern in the EU's trade and environment policies. Thus the EU filed a formal complaint with the GATT against the American secondary embargo on tuna, arguing that nations should not use trade restrictions to influence the regulatory policies of their trading partners in the absence of an international environmental agreement. Yet at the same time, the EU has supported a ban on imports of fur from nations which permit the use of leghold traps, which is vulnerable to the same objection (see also the pesticide case in Chapter 3). Likewise, many of the same criticisms the EU made of the American CAFE standard could also be made of the EU's ecolabelling programme, since both implicitly, though not explicitly, discriminate against imports. Clearly, what does explain these seemingly contradictory positions is that in each case the EU has favoured the interests of European firms.

Finally, as in the case of the EU's own environmental policies, important differences among the Union's member states have emerged. Concerns about animal protection are particularly strong in Great Britain and northern Europe, but are less widely shared in France and the EU's southern states. The Dutch and Austrians have been particularly strong advocates of trade restrictions to protect tropical forests, while the British and Irish remain relatively unenthusiastic about the Union's beef hormone ban (discussed in Vogel 1995). The American secondary tuna embargo directly affected only two member states, Spain and Italy, while the US CAFE penalties only significantly affected German and British manufacturers. And in the case of ecolabelling, many of the EU's southern states have echoed the concerns of many of the Union's trading partners. Nonetheless, to date, the Union has spoken with one voice in international negotiations on each of these issues.

To date, the impact of the GATT/WTO on EU regulatory standards has been modest. No EU regulation has yet to be successfully challenged, though its legtrap ban may be subject to dispute settlement proceedings in the future and its ecolabelling plan has certainly faced considerable scrutiny by its trading partners. In addition, the EU's efforts to restrict tropical hardwood imports and furs from animals caught with legtraps have been constrained by fear of GATT/WTO scrutiny. Still, all these constitute a relatively minor component of EU environmental policy. In general, EU regulatory policies have not yet been significantly constrained by multilateral trade agreements.

It is in part in order to prevent this from occurring that the Union has proposed changes in WTO rules and procedures.

Notes

1 Contrast this view of extraterritorial environmental protection with the one adopted by the EU in the case of pesticide exports (Chapter 3).
2 See 'EU Ecolabelling: Impacts on Trade and Environment,' Case Number 225.
3 Statement of Belinda Collins, Director of the Office of Standards Services, National Institute of Standards and Technology (FNS 1996).
4 Mutual recognition and equivalency approaches are also being considered in the context of EU eco-audit rules. See Taschner (1998).

Bibliography

AE (1996) 'Commission Faces Court Action Concerning Fur Imports,' *Agence Europe*, 16 January.

Banki, Willemijn (1995) 'The European Eco-Label and International Trade: The Compatibility of the European Eco-Label with the GATT' (Brussels: European Environmental Bureau).

BNA (1995) 'Attorney Says Environmental Trends Abroad Will Have Far-Reaching Impact on US Firms,' *BNA International Environment Daily*, 19 October.

—— (1996a) 'EU to Push Ahead with Eco-Labels for Paper, Despite US Objections,' *BNA International Trade Reporter*, Vol. 13, No. 23, 5 June (available on Nexus).

—— (1996b) 'Business Leaders Draft Proposals for US–EU Trans-Atlantic Summit,' *BNA International Trade Reporter*, Vol. 13, No. 22, 29 May (available on Nexus).

Bristow, Paul (1994) 'The European Community Eco-Labelling Scheme,' in *Life-Cycle Management and Trade* (Paris: OECD).

Charnovitz, Steve (1996) 'The WTO Panel on the US Clean Air Act Regulations,' *International Trade Reporter*, 13 March.

Demaret, Paul (1993) 'Environmental Policy and Commercial Policy: The Emergence of Trade-Related Environmental Measures in the External Relations of the European Community,' in M. Maresceau (ed.), *The European Community's Commercial Policy after 1992: The Legal Dimension* (Dordrecht: Kluwer Academic Publishers).

Dunne, Nancy (1993) 'EC Challenge Over US Fuel Economy Tax,' *Financial Times*, 11 May.

EE (1996) 'EU/Canada/US/Russia: Commission Caught in Leg-hold Trap,' *European Environment*, 8 January.

Eiderstrom, E. (1998) 'Ecolabelling: Panacea or Pandora's Box?' in J. Golub (ed.), *New Instruments for Environmental Protection in the EU* (London: Routledge).

ENS (1995) 'Tuna Fishing,' *European News Service*, 24 November.

EW (1994) 'UN Agencies Plan Labelling Scheme for Third World Products,' *Environment Watch Western Europe*, 4 March.

—— (1996a) 'EU Launches Review of Ecolabel Scheme,' *Environment Watch Western Europe*, 16 February.

—— (1996b) 'EU Members Unhappy at Draft Ecolabel Standard,' *Environment Watch Western Europe*, 5 April.

FNS (1996) *Federal Information Systems Corporation, Federal News Service*, US Department of Commerce House Science Committee Subcommittee on Technology, 4 June (available on Nexus).

Fraser, Damian (1992) 'Environment Hit by Too Much Free Trade,' *Financial Times*, 2 July.

Freestone, David (1991) 'European Community Environmental Policy and Law,' in Robin Churchill, Lynda Warren and John Gibson (eds), *Law, Policy and the Environment* (Oxford: Basil Blackwell).

FT (1996) 'EU Warns of Fur Trap Ban,' *Financial Times*, 3 March.

GF (1992) 'Action Urged on Tuna Panel Report,' *GATT Focus*, No. 88, March.

Harding, James (1995) 'Sticking Point for Fresh Green Products,' *Financial Times*, 29 March.

Hegenbart, Barbel (1991) 'Systematics of Interdependencies: Trade Instruments and their Impact on Environment,' in F. Schmidt-Bleek and H. Wohlmeyer (eds), *Trade and the Environment* (Laxenburg, Austria: International Institute for Applied Systems Analysis and the Austrian Association for Agricultural Research).

Housman, Robert and Durwood Zaelke (1992) 'The Collision of the Environment and Trade: The GATT Tuna/Dolphin Decision,' *Environmental Law Reporter*, April.

IUST (1994) 'GATT Rules Against CAFE, Open Door to Conservation Exception,' *Inside US Trade*, 4 October.

Jha, Venna and Simonetta Zarrilli (1994) 'Ecolabelling Initiatives as Potential Barriers to Trade,' in *Life-Cycle Management and Trade* (Paris: OECD).

Kirwin, Joe (1996) 'Eizenstat Sees Room for Improvement in United States' Relations with EU,' *International Trade Reporter*, 13 March.

Lavelle, Marianne (1993) 'Free Trade vs Law,' *National Law Journal*, Vol. 15, No. 30, p. 39, 29 March.

Levinson, Marc (1996) 'Seeing Red Over Green: Why Big Business Hates Ecolabels,' *Newsweek*, 17 June.

Maggs, John (1992) 'EC Will Protect US Tuna Embargo Against 20 Nations,' *Journal of Commerce*, 4 February.

Marshall, Jonathan (1992) 'How Ecology Is Tied to Mexico Trade Pact,' *San Francisco Chronicle*, 25 February.

NWF (1996) 'Profile of the EU Ecolabel Programme,' in *Guarding the Green Choice*, A National Wildlife Federation Trade and Environment Report (Washington, DC: NWF).

Pleydell, Geoffrey (1993) 'Tropical Timber "Bashers" Attacked,' *Financial Times*, 12 May.

PR (1996) 'Commission Launches Guidelines to Promote World Rules on Trade and Environment,' PR Newswire, 28 February (available on Nexus).

PRN (1996a) Public Relations Newswire, 18 February.

—— (1996b) Public Relations Newswire, 2 March.

Rosen, Barry and George Sloane III (1995) 'Environmental Product Standards, Trade and European Consumer Goods Marketing,' *Columbia Journal of World Business*, Vol. 30, No. 1, pp. 75–86.

Simon, Barnard (1991) 'Hunting for a Kinder Kill,' *Financial Times*, 13 December.

Southey, Caroline (1995) 'Brussels Gives Way in Fur Trapping Row,' *Financial Times*, 23 November.

Taschner, K. (1998) 'Environmental Management and Audit Systems,' in J. Golub (ed.), *New Instruments for Environmental Protection in the EU* (London: Routledge).

TE (1993) 'US Auto Fuel-Efficiency Taxes to be Examined by GATT Panel,' *Trade and the Environment: News and Views from the GATT*, 3 June.

UPI (1996) 'EU Sets Deadline on Leg-hold Traps,' UPI Newswire, 4 March.

Vogel, David (1995) *Trading Up: Consumer and Environmental Regulation in a Global Economy* (Cambridge, Mass.: Harvard University Press)

Williams, Francis (1996) 'US Holding Up Green Trade,' *Financial Times*, 26 July.

7

THE WORLD TRADE DIMENSION OF 'GREENING' THE EU'S COMMON AGRICULTURAL POLICY

Andrea Lenschow

This chapter explores the emerging green dimension of the Common Agricultural Policy (CAP), with special attention to its relationship with the international agricultural trading regime. It sheds light on the unusual event of international trade liberalisation pressures pushing forward a 'greening' of European sectoral policy. First, the GATT context operated to impose constraints on the continuation of the CAP, and with it the production of negative environmental side-effects. Second, from the GATT negotiations emerged the positive incentive to shift support for farmers to less trade distorting social and environmental measures.

This chapter begins by investigating the environmental dimension of the CAP prior to the MacSharry reforms and the closure of the Uruguay Round. It describes the – quite limited – integration of environmental concerns into the CAP prior to the 1992 reform and traces its roots to merely internal factors, such as a rising environmental awareness, intra-EC (now EU) trade harmonisation, and the desire to maintain social structures in rural areas, especially in southern Europe. In contrast to other chapters in this volume, the degree and form of 'greening' the CAP at that point was not limited by considerations related to maintaining the international competitiveness of European agriculture. The nevertheless limited character of environmental policy integration into the CAP can be explained, instead, by the firmly institutionalised traditional structure of the CAP, preventing the radical reform necessary to remove the policy's negative environmental externalities, and by the limited funds and political support for expanding structural adjustment measures targeted at environmental objectives.

I will then show how the latest GATT round played an explicitly positive role in the process of environmental policy integration in the CAP during the early 1990s. I argue that the GATT negotiations formed the necessary context for an incomplete and yet dramatic shift in the CAP's objectives and its choice of policy instruments. As the completion of the Uruguay Round foreclosed the option of maintaining the level of the EU's longstanding agricultural support scheme based on guaranteed prices, EU policy instruments to provide support for the farming sectors were partly shifted towards direct income support and service payments, in other words instruments that were not only less trade-distorting but also preferable from an environmental perspective.

A closer look at the GATT negotiations reveals an interesting picture of two-way causalities and policy feedback. Not only did the GATT facilitate an environmental improvement of the CAP, but at the same time, environmental rhetoric helped bring about the political compromise resulting in the completion of the Uruguay Round and the signing of the Marrakesh Accords in 1994. In part, it was the environmental agenda that pointed to the conceptual distinction between trade-distorting and non-distorting agricultural support measures and the applicability of GATT rules only to the former. In an effort to rescue the political deal, however, a certain degree of environmental and social 'window dressing', aimed at minimising the portion of domestic support payments that would be subjected to liberalisation requirements, was employed in the negotiation of operational details of the GATT agriculture agreement.

In short, this chapter argues that in the case of agricultural trade, the liberalisation of the European regime and the provision of GATT exemptions on environmental grounds went hand in hand. It proceeds as follows: the next section focuses on the environmental externalities of the CAP and its limited environmental reforms prior to 1992; it highlights these reforms' exclusively internal rationales and the absence of an international dimension – either negatively or positively. The 'greening' of the CAP through the MacSharry reforms, in contrast, needs to be situated in the international trade negotiations, as explained in the next part of the chapter. The concluding part ventures into hypothetical future scenarios for the CAP and its environmental dimension.

The CAP's environmental dimension prior to 'Uruguay' and 'MacSharry'

Environmental effects of the CAP and early environmental measures

The objectives of the CAP, as set up in 1957 (Arts 38–43 of the Treaty of Rome), did not include environmental protection, nor was the architecture of the CAP compatible with such an objective, as the following decades would show. Most of the steadily rising CAP budget was allocated to the market price support system, based on variable import levies, variable export refunds and intervention prices,[1] which implied a number of economic, social and environmental problems.

In brief, economically the CAP suffered from its producers' insulation from market pressures combined with their unlimited incentive to produce. It resulted in a structural over-supply of several goods – one may recall the European butter mountains and milk lakes. Between 1987 and 1991 the value of EC agricultural surplus increased by a factor of six to approximately 6 billion ECU.[2] The economic costs are borne by the EC budget, hence tax payers, and European consumers.[3] The policy failed equally to achieve its social objectives as the living standard of the average European farmer remains below the population's average and the distribution of payments has been very uneven. The Commission estimated that in the mid-1980s 80 per cent of CAP expenditure reached a mere 20 per cent of European farmers, these being typically the larger and more efficiently operating producers in the northern regions (CEC 1991a: 1).

Environmental degradation in the agricultural sector is the result of a combination of market failures in the agricultural sector, failing to internalise its environmental cost (such as the exploitation and pollution of water and soil resources), and non-market (or government) failures operating to reinforce these market failures (Runge 1993: 96–97, OECD 1994: 110–114). It has been widely acknowledged that the market is biased against environmental protection because the Polluter Pays Principle (PPP) is generally not applied to the European agricultural sector.[4] On the government failure side, the CAP, with its system of guaranteed prices which effectively couples financial support to production output – and surplus – contributes to environmental damage, aside from being trade-distorting, economically costly and socially sub-optimal as just described. The Food and Agriculture Organisation (FAO) has linked the European agricultural production system and practices directly to environmental effects such as the pollution and contamination of soil, water, air and food, due to increased agro-chemical use and livestock effluents, the degradation of natural resources, the disturbance and reduction of biotops and wildlife habitats, and the loss of biological and genetic diversity (cited in Runge 1993: 105).

Despite clear evidence of environment-related market and government failures, reform attempts directed at the European agricultural policy were limited prior to 1992. Although the agro-environmental agenda began to be set in 1985 with the Commission's green paper on perspectives of the CAP which – in response to an intervention by DGXI – proposed that agricultural policy should 'take account of environmental policy, both as regards the control of harmful practices and the promotion of practices friendly to the environment' (CEC 1985a, cited in Baldock and Lowe 1996: 12), the actual changes to the CAP consisted of only marginal modifications of the guidance section of the CAP (structural support for farmers extensifying production, setting aside land and maintaining environmentally sensitive areas) and some quantitative restrictions of agricultural production (the introduction of co-responsibility levies on surplus-producing farmers and a budget ceiling).

Quantitative restrictions were argued to limit the incentive to use agro-chemicals as well as the (over-)production of manure in intensive livestock holdings. The 'Council Regulation 797/85 on Improving the Efficiency of Agricultural Structures' authorised member states to develop national support schemes in environmentally sensitive areas (ESAs). It further offered EC-funded premiums for farmers who engage in extensification and set-aside measures. Two years later, 'Council Regulation 1760/87' was adopted which made ESA support eligible for up to 25 per cent EC co-financing. Following a review of Regulation 1760/87 in 1990, the Commission announced its intention to reinforce the relationship between agriculture and the environment (CEC 1990a) and packaged the three measures (exten-sification, set-aside, ESAs) into the proposal (CEC 1990b) that was to become the core of the 'Council Regulation 2078/92 on Agricultural Measures Compatible with the Requirements of the Protection of the Environment and the Maintenance of the Countryside' which accompanied the MacSharry reforms of the CAP.

The environmental effects of the 1985 and 1987 agricultural structures legislation remained limited, however. They constituted a minimal counter-weight to the environment-unfriendly guarantee section of the CAP. Also, the measures suffered from a northern bias by focusing on side-effects of intensive production and neglecting issues such as soil erosion, desertifica-tion or forest fires which concerned primarily southern member states and parts of France. Consequently, the early set-aside scheme was taken up only by Germany, Belgium, France and Italy; the ESA option was implemented only by the UK, Germany, the Netherlands and Denmark. Only after the possibility of EC financial aid was introduced in 1987 did Italy, France, Luxembourg, Ireland and Spain start small experimental ESA schemes (Baldock and Lowe 1996: 16–17).

Even more limited were attempts to 'control actions harmful to the envi-ronment' in the agricultural sector as declared desirable in the 1985 green paper. The regulatory measures related to the protection of water quality adopted by the EC since the 1970s had little impact on agricultural production practices. The implementation measures generally ignored the prevention principle by calling for remedial action rather than controlling the sources of pollution such as the leaching of agro-chemicals and livestock manure into water resources. In other words, not farmers but water authori-ties were primarily targeted by EC water legislation. Gradual realisation that existing implementation attempts are insufficient to meet quality goals laid down in the EC water legislation has led to recent pressures toward restruc-turing and improving its effectiveness (CEC 1996b).

The 1991 'Council Directive Concerning the Protection of Waters

Against Pollution Caused by Nitrates from Agricultural Sources (91/676/EEC)' has been the first attempt to break with the previous pattern of remedial action by placing the spotlight directly on the farming sector.[5] The Directive does not go as far as requiring the application of the Polluter Pays Principle, however. Nigel Haigh traces its history to the 1980 'Drinking Water Directive (80/778/EEC)' which set a nitrate limit of 50mg/l. But while the Drinking Water Directive focused on remedial action which tends to be costly, the 1991 Nitrate Directive 'aims to reduce and *prevent* the pollution of water caused by the application and storage of inorganic fertilisers and manure on farmland' (Haigh 1992: 4.4–12 and 4.14–1, emphasis added). It obliges member states to develop action programmes to ensure the reduction of nitrate pollution in designated 'vulnerable zones'. Farmers in these vulnerable zones may be compensated by their national governments for the financial losses they incur.

To review, the CAP as originally conceived not only fails to correct negative environmental externalities produced by market mechanisms in the agricultural sectors, it even exacerbates environmental problems through government failures implied in the CAP's price support structure. Since 1985, minimal measures to reduce the negative ecological impact of public intervention were implemented and limited regulatory policies aimed at reducing the market failures introduced. The following section will identify the mostly EC-internal reasons for these changes.

Pressures and constraints related to pre-1992 environmental policy integration

Contrary to other cases discussed in this volume, the initial 'greening' of the CAP since the mid-1980s was not resisted on the basis of its effects on the international competitiveness of European farmers.[6] Simple explanations for this phenomenon are (a) the relative absence of a – domestic or international – environmental agenda that related to European farming practices, and (b) the sheltered existence from international trading pressures led by European farmers, due to CAP protection. Prices received by European farmers for their produce internally and on the export market remained relatively unaffected by international competitive pressures since they were established during the Agriculture Council negotiations which, prior to the completion of the Uruguay Round, took place independently of international agricultural trading agreements.

In tracing early environmental policy integration in EC agricultural policy, it appears that the evolution of the CAP can be explained by reference to primarily internal factors. At the same time, however, internal factors also posed a constraint on more drastic measures. In brief, the increasing salience of environmental issues constituted the motivation for 'greening' the CAP; mounting legal and more notably fiscal challenges to the legitimacy of the traditional CAP created the beginnings of a permissive structure for reforms. On the constraining side, the deep institutionalisation of the CAP in the fabric of the EC prevented radical change.

The salience of environmental issues related to agricultural production emerged more slowly than the realisation and response to negative environmental externalities generated by traditional industrial processes. This was due to the 'green' image of agriculture and the general public's valuation of farmers as 'guardians of the environment' and providers of 'stability and rootedness' (Keeler 1995). The relative blindness with regard to the negative side-effects of the CAP was reinforced by its 'sacred nature' within the Community's framework, the powerful agricultural lobby nationally and in Brussels, and the structural insulation of agricultural policymaking from external pressures such as financial, foreign trade, social or environmental objectives (Lenschow 1995, 1996).

Until the adoption of the Single European Act in 1986, the CAP constituted the core, and hence symbol, of the European integration project; systemic radical reforms of the CAP were therefore out of the question as they would have been perceived as unravelling the larger European edifice (Marsh and Swanney 1980, Tracy 1989, Urwin 1991: 135). The fact that the strong corporatist tradition in European agriculture had been extended to the EC level, with the European agricultural federation COPA and its member organisations deeply linked to the policymaking process, further solidified the status quo in agricultural support. The same was true for the insulated structure of CAP decision-making, characterised by the EC's general functional fragmentation, exacerbated by the separate existence of the Special Committee on Agriculture preparing the Councils. As a consequence, the CAP allowed agricultural ministers and their clientele to escape – already minimal – domestic cross-policy scrutiny, such as pressures from their colleagues with the finance portfolios to avoid budgetary overruns or from environmental ministers to comply with environmental objectives, without subjecting them to at least equivalent scrutiny on the European level. In fact, until the 1988 reforms which introduced a ceiling for CAP budgetary increases, the Agriculture Council acted as irresponsibly as it did quite unchecked.

Several factors began to undermine the CAP's special status in the mid-1980s, however. With food security, the primary objective of the CAP, achieved, the costliness and wastefulness of the policy's operations gained in attention among the European population. This awareness was reinforced by the single market project, which challenged the CAP as the symbolic centre of the EC and emerged as a competitor for scarce European funds. Budget constraints became the primary impetus for subsequent CAP reforms which consequently focused on the problem of finance-draining surplus production. Analysing the policy discourse since the mid-1980s, it appears that the linkage between budget pressures and the environmental problem of

European agriculture has been constructed in two different ways. First, environmental measures which carry the promise of resulting in lower yields due to extensification or reduced input of agro-chemicals are proposed not only as serving environmental interests but also as contributing to surplus reduction. Second, and partly in contrast, to the extent that it could be argued that high budgetary expenditures for the agriculture sector helped support the provision of desirable services such as environmental stewardship in rural areas, a sizeable common agricultural policy could still be justified to an increasingly critical public and hence maintained. On the basis of these two rationales, the 'greening' of the CAP represented an attempt at repackaging a policy under attack into one acceptable to the growing policy community that demanded a say with respect to the CAP's future.

Both 'policy linkages' are contained in a statement by Environment Commissioner Clinton Davis who argued in 1985 that

> [in its] role as the protector of the environment, of the landscape, and of natural habitats [the farming sector] thus *renders services* to society for which there is a real demand. Direct income support, which may be indispensable for income or market reasons and which has the advantage of *not encouraging higher production*, can take account of the role of agriculture in the environment.
>
> (CEC 1985b: 2, emphasis added)

These ideas were quite controversial when first voiced, however. DGVI (agriculture) accepted the notion of rewarding the agricultural sector for rendering environmental services and favoured expanding the structural payments of the CAP in the form of accompanying measures while denying the 'direct link' between the market support and environmental effects, and hence the need to reform the guarantee section of the CAP (CEC 1987).[7] The EP agricultural committee, more radically opposed to the proposed options of environmental integration, criticised the Commission for considering giving support to 'not surplus products but surplus, and rather poor, farmers . . . to do little in rather pretty countryside' (EP 1986: 64). Agriculture ministers were equally sceptical about instituting an EC scheme for rewarding environment-friendly farming, at least to the extent that this would involve a transfer of environmental management tasks to Commission authorities. The fact that the guarantee section remained largely untouched until 1992 (with the exception of some production control measures), and structural measures were expanded only minimally, as indicated above, is therefore not surprising.

The 'story' behind the adoption and elaboration of the structural policy within the CAP aimed at environmentally sensitive areas in 1985 and 1987 is characteristic of the narrow path of opportunities for environmental reform within the CAP. Initially, the ESA scheme was proposed by the

British government for predominantly domestic reasons. As described in more detail by Baldock and Lowe (1996: 13–15), environmental interests placed pressure on the government to limit farming in sensitive areas. While a complex procedure was introduced whereby conservation authorities could limit farming in such areas in return for compensation payments, the ministry for agriculture resisted further-reaching measures under the pretence that under EC law no national aid could be rendered to farmers for purposes other than farming and that therefore a change in EC legislation was necessary to meet the environmentalists' demands. This option was therefore pursued by the UK Government, pushed by environmental groups mounting a European campaign. Despite considerable irritation among other member states and the Commission, the ESA scheme was integrated in Regulation 797/85 even though at that point only as a 'from now on' legitimate option for national aid. In the amended Regulation 1760/87 the ESA payment scheme became eligible for EC co-financing. This brief policy history shows that most agricultural ministers were hesitant to internalise environmental costs or support environmental services domestically. They were equally resistant to changes to the status quo of the CAP that would reduce transfer payment to the farming sector or implied more external control on domestic structures and practices. Consequently, changes in the structure of the CAP needed to be incremental and accompanied by generous compensation for the affected farmers. In a situation where resources were increasingly scarce, and hence the scope for politically acceptable and fiscally possible compensation payments limited, there emerged the political package of combining environmental and economic need, based on the new 'acceptance that supporting farmers to conserve the countryside might also help, albeit in a modest way, to curb overproduction' (Baldock and Lowe 1996: 15).

Recurring environmental rhetoric in subsequent CAP-related Commission documents (CEC 1987, 1988, 1990a, 1991a) underlines the beginning of a change in perspective toward the environment, however. Even though the change in official rhetoric and the perspective of agricultural policymakers may have occurred only for the pragmatic reasons outlined above, in practical terms it produced 'the need to give some substance to the formal commitments made in various policy documents to integrate environmental considerations into agricultural policy' (Baldock and Lowe 1996: 15). Similarly, agricultural policymakers felt increasingly constrained by the publicity given to mounting evidence that the CAP was to a large part responsible for implementation failures of EC environmental legislation, such as the EC's water legislation discussed above. For instance, in April 1990 the World Wide Fund for Nature (WWF) presented vast empirical evidence of environmental degradation caused by agricultural production and called on the Agriculture Council to 'discontinue all subsidies for farming activities which destroy the environment' (AE, No. 5241: 25 April 1990), to encourage environmentally-friendly farming methods, to impose environmental controls, and to introduce the principle of environmental conditionality for financial outlets (discussed in Baldock 1990).[8]

By 1990, the Commission as a whole proclaimed the necessity to make agriculture more responsive to the needs of the environment and proposed not only to extend structural support for the protection of the environment but also to introduce certain forms of environmental conditionality and regulation (CEC 1990a). In a political climate in which environmental issues were considered increasingly salient, the presence of previous legal and declaratory commitments to protect the countryside offered a more and more weighty resource for opponents of the CAP. Reacting to this pressure, even the farming sector began to respond favourably to the environmental agenda by generally favouring agri-environmental support; some representatives of the more marginal elements of the sector (peripheral farmers, extensive or even organic producers) even proposed more radical reforms of the CAP (Lenschow 1996: 364–366).

Indicating that this slow change in attitudes remained far from representing a paradigm shift, until 1992 agro-environmental measures remained largely limited to structural incentives and regulatory measures proved an even bigger challenge in terms of surmounting the opposition from the agro-sector. The already mentioned Nitrate Directive stands apart in that the politicisation of the implementation gap in European water legislation resulted in positive action, although even that turned out to be limited. Aside from legal pressures and the costliness of remedial action, the desire to harmonise emerging member state policy aimed at limiting the application of agro-chemicals and livestock manure[9] as well as the realisation of cross-boundary water pollution, motivated EC nitrate legislation. Delays between the initial Commission proposal in 1988, which had been invited by the informal Environment Council in June of that year (Haigh 1992: 4.14–3), and the adoption of the Directive, were due to member states' attempts to guard their national policies and some disagreement regarding the scope of the problem. In order to satisfy agricultural policymakers and lobbyists, the final EC Directive allows for compensation measures which were justified on the basis that the source of nitrate pollution may not be precisely traceable, either geographically or over time. This again illustrates the limits to radical reforms in the EC agricultural policy.

To sum up, by the early 1990s some linkage between the environment and agriculture was accepted by nearly all interested actors. However, vast differences persisted in the interpretation of the nature of this linkage and hence acceptable policy options. Confronted with two types of market failure – the failure to internalise environmental cost of production, on the one hand, and the failure to price environmental services rendered by the agricultural sector on the other hand – some consensus was emerging to deal with the latter through some form of a remuneration scheme. With regard to internalising environmental costs of agricultural production (i.e. implementing the

Polluter Pays Principle) the debate was highly polarised and remained biased toward the non-enforcement of that principle in regulatory action and the maintenance of public intervention that exacerbated rather than ameliorated the market failures. The 1992 MacSharry reforms were distinct from previous reform rounds in their more systematic attempt to deal with the government failure implied in the CAP, even though the need to 'pay off' farmers for any concessions on their part continued to shape the drafting of the 1992 MacSharry reforms. Significant from the perspective of this volume is the fact that the most radical environmental *and* trade liberalising reforms of the CAP to date took place in the context of the Uruguay Round. Before turning to this connection, let me briefly summarise the outcome of the MacSharry package.

The MacSharry reforms – an environmental milestone?

The MacSharry reforms that were adopted in 1992 resulted in a package that from an environmental point of view reduced the magnitude of the government failure and expanded measures to pay farmers for the provision of public services which would remain unremunerated under pure market conditions. The reform package introduced the most drastic price cuts to date (primarily in the cereal sector) in exchange for direct compensation payments to those farmers who agreed to 'set-aside' 15 per cent of their arable land. Price support for beef was equally reduced and farmers compensated through headage payments for up to a certain number of cattle per acre. In other words, the price guarantee system – coupled to production – continued at a lower level of support and was now supplemented by a new system of direct income support, offering compensation for decreasing price levels – mostly de-coupled from production. In addition, accompanying measures were introduced, in part to reward farmers for providing a public service in the form of preserving the rural environment (i.e. to address the second market failure), but also as a financial inducement to farmers to reduce the environmental costs they are inflicting through their production activities (i.e. to pay polluters for polluting less) and, of course, to address the surplus problem. To quote the Commission:

> an agri-environmental action programme was adopted to give recognition to the dual role of farmers as producers and as stewards of the countryside, and to encourage farming practices which are less intensive and more in tune with environmental constraints . . . which should also make a positive contribution to rebalancing markets.
>
> (CEC 1991b: 2)

The accompanying measures amounted to more than 5 per cent of the guarantee sections of the 1995 and 1996 CAP budgets, with no equivalent transfers existing prior to the 1992 reform. On the basis of the EU budget the precise effect of the cuts in the price support mechanism on the volume of financial transfers cannot be determined, due to significant intervening factors like fluctuations in world agricultural prices. Considering the continuing growth of the CAP budget, however, it seems safe to conclude that these cuts were in large part offset by the – environmentally more friendly – income support payments, including the accompanying measures.[10]

Table 7.1 summarises the achievements of the MacSharry reform for the environment from the perspective of market and government failures, policy alternatives and actual choices.

The table reveals that the MacSharry reforms took steps to correct both market and government failures, without, however, eliminating either. In view of previous attempts to reform the CAP and the enormous obstacles encountered in the form of firmly institutionalised policy structures protected by powerful defenders of the traditional CAP, the following part of

Table 7.1 Environmental dimension of the CAP before and after the 1992 reforms

Market failures	*Public policy – 1992*	*Reform options*	*1992 Choices*
Failure to internalise environmental cost of production	Exacerbate market failure through protectionist system coupled to production output (government failure)	• status quo	
		• reduce government failure (a) mixing old system and direct income support	X
		(b) switching to direct income support	
		(c) ending protection	
		• reduce market failure (a) regulatory measures	X
		(b) economic instruments	X
		(c) voluntary agreements	
Failure to price and remunerate environmental services	Very limited structural assistance to producers in environmentally sensitive areas	• status quo	
		• widen structural assistance (a) EU level	X
		(b) (sub-)national level	X
		• market reform	

this chapter looks into the conditions under which such 'milestone' reform could be adopted.

The world trade context

The MacSharry reforms may not represent an optimal policy improvement for the environment, but they represent an agreement that lies above the lowest common denominator of interests voiced by agricultural ministers and their clientele at the beginning of the reform negotiations.[11] Powerful agricultural interest groups and their 'agents' in government and administration, who had successfully defended the CAP in its original form in the past, conceded comparatively substantial price cuts and a partial switch to income support. Ministers of agriculture agreed, despite previously voiced scepticism, to an expansion of EU-administered agri-environmental assistance. I argue that the 'environmental' reforms were facilitated by the political framework provided by the ongoing GATT negotiations at the time. In short, the 1992 CAP reforms seemed pressing for a variety of serious internal reasons ranging from economic inefficiencies, socio-economic decline of rural areas, and the CAP's ever-rising burden on the EU budget, to its negative environmental impact. However, these internal pressures had existed since the 1970s. Although reform pressure was continuously mounting with the rise in membership, new demands to be met with European budgetary means (most notably the structural funds) and the EC's legal commitment to integrate environmental concerns into other policy areas, reform attempts in the 1980s had nevertheless left the problematic structure of the CAP entirely intact, and environmental modifications were of limited impact, as described above.

Decisive for the reform in the end was that, by the late 1980s, agricultural trade had been irremovably established on the agenda of the ongoing Uruguay Round and the EC was put under increasing pressure to dismantle the trade-distorting elements of the CAP.[12] Before focusing on the Uruguay–environment link, a few words are needed about the general issue of agricultural protectionism in previous GATT rounds and the nature of the Agriculture Agreement adopted at the conclusion of the Uruguay Round.

Agricultural trade liberalisation in GATT/WTO

The CAP support system was established on the basis of previous national policies, with all six original EC members (with the possible exception of the Netherlands) being used to pursuing agricultural policies based on market price support. It reflected the view that world markets were unreliable and tended to be dominated by the US, undermining the European goals of self-sufficiency in foodstuff and the protection of a rural sector that was based mostly on family farms. The establishment of the CAP benefited from an initially tolerant reaction by the US, due to its 'generally benevolent attitude to the emerging Community on wider political and economic grounds' (CEC 1994a: 63). Trade disputes with the US began in the 1960s, but the EC succeeded in resisting agricultural trade liberalisation measures in previous GATT rounds, claiming that 'the CAP was still in its formative stage and could therefore not be the subject of international negotiations' (Buckwell 1991: 235).

The CAP prompted negotiations within the GATT[13] for the first time in the Dillon Round (1960–1961), resulting in a binding of duties on oilseeds, oilseed products and cereal substitutes at low or even zero levels and the external tariff on sheep meat at 20 per cent. These early agreements were viewed as serious constraints in later years when farmers continued to rely on cheap, imported cereal substitutes for livestock feed instead of using the more expensive, domestically produced surplus cereal. From an environmental point of view, the reliance on imported cereal substitutes contributed to a concentration of livestock – and manure – production near ports. The US and the Community clashed seriously on agricultural matters during the Kennedy Round (1964–1967) where the EC proposed a system of concerted market organisation and market sharing among major agricultural exporters and a common denominator for all agricultural support. The EC's negotiating partners rejected the proposal on the grounds of the limited reduction in agricultural protection and the intervention into domestic agricultural policies they implied. Consequently, the negotiation results remained modest and posed no challenge to the CAP. The Tokyo Round (1973–1979) produced significant results for world trade in general, but tariff reductions in the agricultural sector were limited to tropical products in addition to international agreements concerning the beef and the dairy sector.

Agricultural trade liberalisation eventually became a core element on the agenda of negotiators during the Uruguay Round (1986–1994) with the objective 'to improve the discipline and predictability of world agricultural trade by correcting and preventing restrictions and distortions, especially those linked to structural surpluses' (CEC 1995: 14). Gradual but significant reductions in agricultural support and protection measures were intended in the areas of internal support policies, export subsidies and market access. By joining the 1986 Punta del Este Declaration launching the Uruguay Round and committing the negotiators to 'increase discipline on the use of all direct and indirect subsidies and other measures affecting directly or indirectly agricultural trade' (Ingersent et al. 1994a: 60, quoting the declaration), the EC explicitly and for the first time exposed the CAP's principles and mechanisms to international negotiation. Several observers conclude that the EC's more open negotiating stands during the Uruguay Round can be traced to a conscious attempt by the Commission to impose on the domestic political process some constraints from outside the narrow agricultural arena, [with the consequence that the] agriculture ministers felt unable to oppose the general will to have a successful Uruguay Round, and were prepared to go along with the possibility of some external constraints on their action rather than risk being blamed for heightened trade tensions.

(Moyer and Josling 1990: 190)[14]

In other words, the Uruguay negotiations placed agricultural policymakers in the EC in a broader political framework which challenged their insulated operations. By effectively linking the future of the CAP to an agreement involving not only international actors but also representatives of other policy sectors, its insular existence responsible for its prior evolution along a narrow path (ending in a cul de sac) was opened, and alternative policy paths emerged on the horizon.

In the final agreement,[15] export subsidies are to be reduced by 21 per cent by volume and 36 per cent *ad valorem*, and restrictions to market access (the Community's variable import levies and export subsidies) will be turned into fixed customs duties and reduced by 36 per cent over six years (with a minimum of 15 per cent per product). The variable aspect in the previous Community system is maintained, however, through a safeguard clause under which additional duties may be applied if the import volume exceeds a specific threshold level or import prices fall below a certain level. More significant for the environmental dimension of the deal, and hence the focus of this chapter, domestic support for agriculture was agreed to drop by 20 per cent compared with the 1986–1988 reference period, with those forms of support that are designated to fall in the so-called 'green box' (aid that has no effect on trade or production) being exempted from the reduction commitments. Both the direct income support scheme that was devised to compensate farmers for cuts in price guarantees and the accompanying measures were declared to meet the 'green box' criteria, despite some doubts regarding the true degree of decoupling and non-trade distortion of the compensatory measures in particular (more below).

With some last-minute adjustments, the GATT Agreement on Agriculture was signed by EU members in the conviction that the post-1992 CAP was compatible with the agreement. Commissioner René Steichen concluded in late 1994 that

the Community will be able to meet its new commitments under GATT without having to impose further constraints on farmers . . . [and that] with the outcome of the GATT negotiations, the Community has been able to attain two fundamental goals:

- it has contributed toward constructing an international trading regime which is more market-oriented, without giving up Community preference;
- it has ensured that its international commitments are compatible with the CAP reform, retaining sufficient flexibility to be able to manage by itself the schemes which are at the base of the reformed CAP, such as direct aid and rural development.

(Steichen 1994: 4)

The CAP–GATT debate and the environment link

The Commission summarised the basic action parameters of the Community during the GATT negotiations as follows:

> as the foremost world agricultural trader, the Commission, by changing its rules, is stating its willingness to join the movement towards freer trade advocated at international level while preserving the basic principles and instruments of the CAP (CEC 1992: 4).

In insisting basically on the maintenance of the status quo of the CAP, the EC's original bargaining position seemed as unmovable as in previous years. However, even though only a few member states whole-heartedly favoured agricultural trade liberalisation (the UK, the Netherlands, Denmark), most members, including the European Commission, were becoming worried about disruptions in international trade in general – hence their willingness to contemplate the necessary CAP reforms increased. Most notably, in Germany, with its considerable interest in a liberal trading regime for industrial products, preparedness 'to reform the CAP was closely linked with the need to secure a successful completion of the GATT round. It seemed clear that without a compromise on CAP there would be no positive end of the GATT negotiations' (Hendriks 1994: 66).

Commissioner MacSharry exploited the emergence of an increasingly explicit conflict of interests between trade and agricultural policy objectives within the member states in his discussions with the Agriculture Council. However, he remained constrained by the continuing power of the agriculture representatives domestically and in Brussels and the fact that the kind of reforms individual members were willing to contemplate varied widely in scope and direction. In short, even though MacSharry was able to apply pressure on the Council by arguing that the Community's hands were tied by the larger GATT context (Lenschow 1996: 374), the possible margin of departure from the traditional structure of the CAP remained limited. The CAP continued to be the 'classic story . . . of how, once decisions have been made and structures established, inertia sets in to make any reform extremely difficult' (Urwin 1991: 185). Consequently, the reform strategy focused on cutting the directly *trade-distorting* features of the CAP within a politically acceptable margin and compensating farmers for their losses through measures acceptable under GATT rules.

The distinction between trade-distorting and non-distorting measures corresponds to the legal and political reality of the GATT, governing only international trade relationships and having no authority over domestic policies other than those with trade-distorting effects. Long-time CAP analyst Harvey stresses that

> the achievable objective of multilateral negotiations is, therefore, to minimise trade distortions, not necessarily to eliminate protection or domestic income support. . . . It is now recognised that domestic support of agriculture must be allowed to continue within this constraint, given national desire so to do.
>
> (Harvey 1994: 237)

Already the 1987 EC position paper had referred to the possibility of compensating producers for reduced price support with *decoupled payments*, that is, 'direct methods of supporting farmers' incomes which are not linked to output' (Ingersent *et al.* 1994a: 61–62). On this principle there existed no disagreement with the main GATT negotiating partners. The US, in its revised position paper on agriculture of 15 October 1990, elaborated on the distinction between more or less trade-distorting domestic support measures. It proposed that most trade-distorting domestic support measures, such as market price supports, deficiency payments and production-linked input subsidies, be cut substantially, while non-trade-distorting agricultural programmes, such as environmental protection, resource retirement and diversion, income safety net programmes and *bona fide* food aid, be exempt from any reduction commitment (ibid.: 69).

The challenge to distinguish clearly trade-distorting domestic support measures from others was therefore not one of principle but one of operational detail. Multilateral trade negotiators needed to identify criteria for distinguishing between agricultural protection and trade distortion. In the absence of such – mutually agreeable – criteria, the GATT negotiations showed that the presence of primarily domestic social and environmental objectives of a policy helped diminish its GATT conflict potential even in the case of trade-distorting effects.

Least controversial in this respect were the accompanying environmental measures adopted under the MacSharry reforms, because they represent payments for the provision of a public good with minimal effects on agricultural trade. Free trade purists may argue that some environmental programmes support the production of marketable goods that otherwise would not have been produced and hence distort trade, but the public good character of the measures is likely to outweigh the distorting effects.

It is less clear whether the new direct income support payments (compensating for price support cuts) introduced in the MacSharry reforms ought to qualify as non- or minimally trade-distorting measures, and hence GATT exemption, that is, 'green box' treatment. First, EU compensation payments have actually been made conditional upon continued sowing on the farmers' permitted crop area (historic area minus set-aside). Therefore, the payments are not fully decoupled from production, even though they do limit compensation to historic levels of production. Second, similar to the distorting effect described above,

> to the extent that [compensation payments] enable resources to remain in agriculture rather than be encouraged to leave, as they would be under genuine and uncompensated free trade . . . the agricultural sector will be larger than without the compensation payments and hence remain distorted compared with free trade.
>
> (Harvey 1994: 246–247)

Mandatory set-aside imposed by the MacSharry reforms on compensation recipients would not be necessary under a fully decoupled regime. They were in fact 'included in the CAP reform programme in large part because the price level was not reduced far enough' (Josling 1994: 518).

The fact that the EC compensation schemes as well as US direct income payments were accorded 'green box' status represents a political compromise that contributed to the rescuing of the Uruguay Round. In pursuing this compromise, EC negotiators believed that 'the environmental aspect of the reform will facilitate its acceptance outside the EC' (UK Minister of Agriculture, John Gummer, quoted in AE, No. 5670: 30 April 1992). Taking advantage of the emerging environmental agenda in international trading fora, agriculture ministers were confident that the new compensatory aids would not be subject to any disciplines arising from a GATT settlement. And indeed, the GATT 'Agreement on Agriculture' adopted the argumentation of the EC in 'its commitment to the liberalisation of trade through reducing domestic support for agricultural production, particularly of production-linked agricultural subsidies' and the creation of exemption 'boxes' for measures 'that have, at most, a minimal impact on trade', and a potentially positive one on output reduction and the environment (GATT 1994).

Sceptics may argue that the lack of clearly defined criteria qualifying domestic support measures for GATT exemption will turn the 'green box' into 'a repository of all policies that countries wish to shelter from international attack' (Josling 1994: 520). Equally, the reform's surplus-reducing and environmental 'selling points' such as the mandatory set-aside are not only second-best from a surplus point of view but also ambiguous with respect to their environmental benefits because the scheme opens the possibility of 'slippage' in that farmers may set aside their least productive area and shift agro-inputs to the land promising higher yields. But nevertheless, the reformed CAP can be characterised as 'qualitatively better than before' from a liberal trading and an environmental perspective (ibid.: 519).

It appears that the exogenous pressure of the GATT negotiations accelerated the previously gradual reframing of the CAP from an 'industrial' to a 'rural development' policy (with its social and environmental components) for which the agenda was set by the Commission in the mid-1980s. While considerable misgivings continue to exist within the European farming population (represented in Brussels by COPA) with respect to the shift to direct income aid, even farming associations began to accept the role of the farmer in the protection of the rural environment and criticised the accompanying measures solely for their limited nature that failed to reward the sector adequately for its provision of public goods. COPA has equally refined its position with respect to multilateral trade agreements, switching from a merely defensive position to calling on the GATT/WTO to recognise more fully the social and environmental dimension of agriculture. Only recently COPA argued that '[I]n world trade with agricultural products, it is important to define competition conditions in order to avoid destabilising environmentally friendly production systems, and to facilitate sustainable development in all regions throughout the world' (COPA/COGECA 1995a: 3). The following, concluding part will point to potential problems implied in this seemingly increasingly cosy relationship between agricultural interests, environmentalists and free traders.

Discussion and scenarios

The discussion so far has demonstrated that the agri-environmental agenda served as a bridge between defenders of the CAP and proponents of trade liberalisation. Both camps were able to reach face-saving agreements, in part due to the casting of continuing extensive payments to the farming sector as direct or indirect contributions to environmental protection. In this final part, I will argue that the compromise reached is insufficient and even unstable from the perspective of environmental and trade liberalisation objectives alike.[16] Already mentioned in the previous section were doubts with respect to the non-trade-distorting characteristics of the environmental and income compensation measures introduced in the MacSharry reforms. Second, as already hinted in Table 7.1, trade liberalisation is not sufficient to deal with the negative environmental externalities of agricultural production. Third, the adoption of environmental rhetoric by the farming sector has great potential to be abused in future attempts to block rather than extend trade liberalisation, as is already becoming evident in EU relations with its neighbours in Central and Eastern Europe (CEEC).

Economic and environmental insufficiencies of the CAP–GATT deal

While the 1992 CAP reform has been presented as a dramatic departure from past practice, doubts persist whether it will actually result in the anticipated economic, fiscal, social and environmental effects. Pressure on the EU budget certainly has not dropped, causing concern with respect to the planned enlargement toward CEEC and the possible need for further reforms. Most critical observers of the CAP[17] foresee the necessity of a(nother) radical CAP reform to accommodate the CEEC. In contrast, most national delegations to the Agriculture Council advocate continuity of the CAP and place the responsibility for adjustment on the prospective new members, arguing for a long transition period. Similarly, the Commission is not contemplating radical changes.[18] Past experience suggests that for the CEEC expansion to trigger deeper cuts in the trade-distorting and costly price support mechanism, additional exogenous pressures need to be exerted. But, already limiting the space for external attack, Commissioner Fischler recently called upon

> the US government to understand that the European Agreements cannot be dealt with in GATT/WTO as a simple case of free trade area, as some US authorities seem to believe, for these agreements are simply a stage towards an enlarged EU and a wide area of stability in Europe.
>
> (AE, No. 6674: 24 February 1996: 11)

In other words, current rhetoric suggests little inclination among CAP policy makers to go further along the path of trade liberalisation.[19]

Turning to the environmental improvements accomplished with the MacSharry reforms, scepticism prevails as well. While Agriculture Commissioners Steichen and Fischler reported a notable drop in the use of chemical fertilisers and pesticides as a result of the partial decoupling of assistance from production (Steichen 1994: 3, Fischler quoted in AE No. 6658: 2 February 1996: 8), an expert report commissioned by the Directorate General for Economic and Financial Affairs concludes that

> [T]he current CAP market reforms are unlikely to change dramatically the level of externalities, both positive and negative, produced by EC agriculture, although there should be a better balance towards more extensive crop and livestock production.
>
> (CEC 1994a: 29)

These contrasting views can be traced to the analysts' different focus on market and government failures. The price cut and 'decoupling' policy may indeed work towards reducing the government failure which exacerbates the failure of the market to integrate environmental cost, but it fails to tackle the existing market failures, for instance through additional regulatory measures or economic instruments (such as input taxes).[20]

> Liberalisation of agricultural policies would involve a change in incentive structures that . . . is likely to reduce both the damage inflicted on the global environment and the chemical residues in the food produced by the world's farmers. Both the environment and welfare could be enhanced even further if the removal of distortions to the relative price of food products were to be accompanied by the introduction of optimal environmental policy instruments and/or the removal of distortions to farm input prices.
>
> (Anderson 1993: 167)

However, even the effective reduction of the environmental non-market, government failures of the CAP itself is in doubt due to the indirect mechanism linking changes of market support to environmental results combined with the absence of monitoring and control provisions or a form of environmental conditionality. This oversight is due to the policy's attempt to achieve various goals (surplus reduction and environmental protection) with one instrument (price cuts). Many policy analysts argue that domestic agricultural and trade targets in agriculture should be matched with domestic agricultural and trade instruments, and environmental targets with environmental instruments (Baldock *et al.* 1992, Runge 1994, Floyd 1995). Their advice overlooks, however, that 'policy analysis is one thing; politics another', and especially in agricultural matters (Grant 1995: 17).

The absence of monitoring and control elements or environmental conditionality equally undermines the effectiveness of direct environmental projects. In addition, they suffer from limited funding and the vague and nearly all-encompassing legal parameters allowing projects of quite dubious environmental merits to be funded. Council Regulation (EEC) 2087/92 calls upon member states to develop and financially support (sub-)national agro-environmental programmes, but the national responses vary dramatically.[21] Environmental organisations have pointed out a lack of integration of the accompanying measures with other CAP and structural policies (Birdlife 1994). For instance, farmers may be discouraged from long-term environmental set-aside as long as they are also mandated to set aside land in order to qualify for compensation payments; payments for environmental projects may not suffice as an incentive to forgo continuing market support.

Despite much environmental rhetoric, even agriculture ministers were aware of the limited environmental benefits to be expected from the reforms. UK Agriculture Minister Gummer announced that the fuller integration of environmental protection in the CAP constituted one of his priorities during the British presidency in the second half of 1992 (AE, No. 5772: 15 July 1992) and he invited the Commission to put forward proposals to this effect (AE, No. 5826: 1 October 1992). Nothing much has happened since then, as the subsidiarity and deregulation debates have blocked an expansion or redirection of EU activities without yet leading to effective measures on the national level (CEC 1996a).

A renationalisation scenario?

Given the constraints on continuing radical changes of the CAP and environmental policy integration, the question arises whether the − at least partial − renationalisation of agricultural policy may offer a solution to the environmental *problématique*. Such a move would correspond to the EU's increasing attempt to implement the subsidiarity principle and avoid overregulation from the European level. Indeed, the at least partial devolution of environmental and general income assistance programmes to the national level may be a sensible way of dealing with highly divergent regional climatic and geological conditions as well as socio-economic and political preferences (Ockenden and Franklin 1995). Wilkinson stresses that Council Regulation No. 2078/92 has increased the scope for member states to support the income of their farmers and views this as a path to pursue in the future (Wilkinson 1994: 26–29).

There may be limits, however, to a renationalisation strategy − politically and with respect to its environmental benefits. Given the experience that (partial) 'renationalising' has been and continues to be a taboo in European agricultural circles,[22] the introduction of this concept with respect to the implementation of the guarantee and guidance sections of the CAP borders on revolutionary. Commissioner Fischler's recent remarks on the subject have triggered a defensive reaction in the Council with respect to the 'common' guarantees of the CAP. For instance, agriculture ministers warned that the prospect of Eastern enlargement must not 'serve as a pretext for dismantling or *renationalising* the CAP' (AE, No. 6617: 1 December 1995). Partial renationalisation of the guidance section, on the other hand, may be politically possible, considering the member states' past reluctance to allow EU structural intervention and the relative costliness of EU management of mostly sub-national programmes (Whitby 1996). However, existing options for national agri-environmental measures suggest a danger of imposing too few environmental safeguards.[23]

In the absence of EU controls and even limited co-ordination, renationalisation may also reintroduce EU internal trade distortions, allowing a situation where the richer countries will support their farmers for simply maintaining their presumably environment-friendly production methods while poorer countries may not be able to afford similar assistance to their rural sector. From an international trade perspective, renationalisation does not alter the EU's obligations. In other words, income support measures do not escape GATT scrutiny by being devolved from the European to the national or sub-national level. Such measures may make EU internal coordination during the next WTO round more cumbersome, however. In sum, while the renationalisation of agri-environmental measures may contribute to relieving the EU budget and even the protection of the environment, adequate framework conditions will need to be established on the EU level to ensure positive environmental effects, prevent trade distortion within the Union, and allow for effective EU representation during future trade negotiations.

A green protectionist scenario?

The EU has been at the forefront of actors arguing the compatibility of environmental protection, economic development and competitiveness in a free trade regime (CEC 1993, 1994b, 1996c, Advisory Group on Competitiveness 1995). With respect to agriculture it is stating its 'keen interest in ensuring that WTO members respect their commitments . . . and in particular those regarding the reduction of subsidised exports and market access' (CEC 1996d: 8). Nevertheless, there are indications that a proper balance between maintaining a favourable framework for environmental protection and agricultural trade liberalisation may not be easy to find.

'Green protectionism' may indeed become a consequence of the new image adopted by the agricultural sector. The prospect of bilateral trading agreements with agricultural competitors (e.g. South Africa) has already caused opponents to these agreements within the EU to point to the higher environmental costs borne by European farmers and hence the competitive disadvantage suffered. Agriculture ministers have expressed

> alarm to see the Union taking giant steps on the attractive but overly risky path of free trade at regional level. . . . [Admitting that] liberalising of trade, which will doubtless continue, can offer benefits for the agricultural community [they emphasised that . . .] given the social, environmental and animal protection constraints which are stricter in the Union than in a good many third countries, free trade can offer an unfair advantage to its partners.
>
> (AE, No. 6727: 13/14 May 1996: 11)

Also, the prospective enlargement of the Union towards CEEC has caused alarm in the agricultural community. As integration of CEEC is likely to introduce new pressures for reform on the CAP, COPA is preparing its 'defence' by pointing to the enormous 'divergence between the economic, social and environmental conditions in the EU and the candidate countries', arguing that the 'preparations for enlargement must therefore be thorough and must not be rushed' (COPA/COGECA 1995b). The prospect of 'green protectionism' should not be overstated, however, as presently the transformation in the transfer mechanisms to the farming sector from price guarantees to 'green' income support is still limited.

To conclude, as the status quo and these slowly crystallising future scenarios indicate, the political debate and reality concerning the compatibility of environmental objectives and trade liberalisation, and the possibility of green protectionism, as well as the future for environmental policy integration into European agricultural policy, have not yet been decided. The current IGC has not been a milestone in that evolution, failing, for instance, to agree on a rewording of Article 39 of the Treaty and failing also to introduce the protection of the environment as one of the CAP's policy objectives.[24] Given the remarkable stability of the CAP in the past, the next turning point may have to wait until the next WTO multilateral trade negotiations, which may then take place in the framework of new WTO–environmental parameters. Certainly in the past two decades, the context of external pressure has stimulated CAP reform more effectively than even extreme internal crises, ranging from enlargement to a budgetary crunch.

Notes

1 Import levies raise the price of the imported good above a so-called threshold price which lies above the Community price level. Variable export refunds compensate EC exporters for the difference that exists between Community and world market prices. Finally, in the event that EC prices fall beneath the intervention price, Community authorities purchase and store EC farm products in order to reduce available supply and hence raise market prices.

2 These data are weighted by 1979 world market prices (CEC 1994a: 15).

3 In 1992, CAP spending amounted to 35.8 billion ECU (Tracy 1993: 172). Transfer to the agricultural sector including public spending by national governments was 51.8 billion ECU, while the cost of agricultural support borne by consumers has been calculated as 69.3 billion ECU. The share of the GDP of total agricultural policy transfers in the EC equaled 2 per cent (CEC 1994a: 26).

4 The most systematic study in this context has been produced by Baldock and Bennett (1991). They argue that agriculture may claim to be a special case with respect to the Polluter Pays Principle (PPP), due to the predominance of non-point sources of pollution which are difficult to monitor and control, complex cause–effect relationships in some agro-ecosystems, social and political considerations connected to small farmers especially, and limits to a radical change in land use patterns. Nevertheless, they regret that 'no country has made a systematic effort to apply the PPP to agriculture' (ibid.: 11) and argue that ' "second best" solutions – involving regulations and, perhaps, subsidies – will in practice be necessary if agricultural pollution is to be effectively controlled' (ibid.: 230).

5 It is summarised in detail by Reeve (1994: 103).

6 With respect to the Nitrate Directive it can be argued that the existence of international environmental agreements or action plans on the protection of the North and Baltic Seas induced the agro-environmental regulatory measure.

7 Ignoring that the majority of EC price support went to the richer, larger, modernised, northern farmers (cf. CEC 1991a) who were induced by production incentives to increase their use of agro-chemicals and to employ the most intensive and productive farming methods, DGVI argued that the effect of production incentives is indirect, and hence in its aggregate inconclusive, as other (more marginal) farmers contribute to the protection of the environment due to the help they receive to stay in existence (CEC 1987: 7).

8 The WWF analysis and demands were echoed in the European Environmental Bureau's (EEB) semi-annual memorandum to the Council Presidency (AE, No. 5309: 3 August 1990).

9 In other words, not concerns with international competitiveness but the issue of intra-EC competition contributed to the adoption of the Nitrate Directive. Similarly, measures to regulate pesticide production and use were in part inspired by the desire to harmonise European market conditions (Lenschow 1996). Further, the permissible scope of national aid for structurally disadvantaged regions or individual farms has been an issue between EC agricultural ministers fighting to protect the interests of 'their' farmers.

10 While the offsetting process does not apply to all affected farmers equally, the initially intended redistributive effect in the shift in policy mechanisms was largely diverted in the final design of the CAP reform which theoretically permitted compensation for mandatory set-aside for all farmers.

11 It is, of course, not surprising that the reforms were less dramatic than the initial Commission proposal.

12 In 1990–1991 agriculture contributed only about 8.5 per cent to the total of EC exports and accounted for 11.5 per cent of Community imports. However, this implied that the Community had become the second largest agricultural exporter after the US (mostly in cereal, dairy products, beef, wine and sugar) (CEC 1994a: 23).

13 This synopsis of pre-Uruguay GATT rounds is based on Tracy (1989: 347–351), Ingersent et al. (1994a: 56–58), and CEC (1995: 9–11).

14 See also Ingersent et al. (1994a: 61), Josling (1994: 514–515) and Baldock and Lowe (1996: 11).

15 This summary of the agricultural provisions agreed to in the Uruguay Round's 'final act' is based on Ingersent et al. (1994b) and CEC (1995: 22–30).

16 The domestic socio-economic implications of the CAP reforms are not the focus of this discussion, though certainly crucial for future policy planning.

17 These include the Swedish and possibly the British governments among the member state delegations.

18 Commissioner Fischler recently presented the Commission's position in favour of pursuing future CAP reform on the basis of the MacSharry package, focusing on its operational simplification, the application of the subsidiarity principle, and the definition of a clearer distinction between market policy and income support. In that context, income support will be more closely tied to social and environmental objectives. (AE, No. 6615, 29 November 1995 and No. 6617, 1 December 1995).

19 To the extent that the Commission succeeds in clarifying the distinction of market and income support and the environmental dimension in these categories of support, it might diminish its chances for 'window dressing' in future trade negotiations, however.

20 Regulatory measures in particular suffer from the problem of identifying the polluter in agricultural production. Water pollution, the main problem arising from European agriculture, occurs with considerable time lag from the actual

application of polluting substances, and the precise tracing of the source is quite impossible. A comprehensive policy to deal with water pollution and exploitation is only now in preparation (CEC 1996c).

21 Ninety per cent of all Austrian farms and 50 per cent of all German farms participate in the scheme, whereas (for different reasons) less than 10 per cent of Belgian, Danish, Italian or Dutch farms do (data cited in Brouwer and van Berkum 1996: 10).

22 In reality, the 'common market' character of the CAP has begun to disappear since the introduction of the Monetary Compensation Amounts in the early 1970s.

23 Ockenden and Franklin quote claims that Germany has used the discretion provided within the ESA Regulation to provide subsidies to farmers (1995: 45). See also Birdlife (1994).

24 For such a proposal see WWF (1996).

Bibliography

Advisory Group on Competitiveness (1995) 'Improving European Competitiveness,' second report, reproduced in *AE – Europe Documents*, No. 1966/1967, 14 December.

AE (various issues 1990–present) *Agence Europe*, General News (Luxembourg, Brussels: Agence Internationale d'Information pour la Presse).

Anderson, Kym (1993) 'Agricultural Trade Liberalisation and the Environment: A Global Perspective,' *The World Economy*, Vol. 15, No. 1, pp. 135–152.

Baldock, David (1990) *Agriculture and Habitat Loss in Europe* (Gland: WWF).

Baldock, David and Graham Bennett (1991) *Agriculture and the Polluter Pays Principle. A Study of Six EC Countries* (London and Arnhem: Institute for European Environmental Policy).

Baldock, David and Philip Lowe (1996) 'The Development of European Agri-environment Policy,' in Martin Whitby (ed.), *The European Environment and CAP Reform. Policies and Prospects for Conservation* (Walingford: CAP International), pp. 8–25.

Baldock, David, Nellie Bandarra and Guy Beaufoy (1992) *The Implementation of Agri-Environmental Policies in the EC* (London: IEEP).

Birdlife (1994) Birdlife International European Agriculture Task Force, *Implementation of EU Agri-environment Regulation 2078* (Cambridge: Birdlife International).

Brouwer, Floor and Siemen van Berkum (1996) 'CAP and Environment in the European Union,' paper prepared for the Conference on European Agriculture at the Crossroads: Competition and Sustainability at the University of Crete, 9–12 May.

Buckwell, Allan (1991) 'The CAP and World Trade,' in Christopher Ritson and David Harvey (eds), *The Common Agricultural Policy and the World Economy* (Wallingford: CAB International), pp. 223–240.

CEC (1985a) *Perspectives for the Common Agricultural Policy – The Green Paper of the Commission* (Luxembourg: Office for Official Publication of the European Communities).

—— (1985b) 'Agriculture and the Environment,' Intervention by Commissioner Stanley Clinton Davis at the Hearing of the European Parliament's Environment Committee on 16 September 1985 (Brussels: CEC, Spokesman's Service).

—— (1987) Commission of the European Community, Directorate-General for Agriculture, 'Environment and the CAP,' *Green Europe. Newsletter on the Common Agricultural Policy*, No. 219 (Luxembourg: Office for Official Publications of the European Communities).

—— (1988) 'Commission Communication: Environment and Agriculture,' COM(88)338 final (Luxembourg: Office for Official Publications of the European Communities).

—— (1990a) 'Agriculture and the Environment,' Press Release, P54, 25 July (Brussels: CEC, Spokesman's Service).

—— (1990b) 'Proposal for a Council Regulation (EEC) on the Introduction and the Maintenance of Agricultural Production Methods Compatible with the Requirements of the Protection of the Environment and the Maintenance of the Countryside,' COM(90)366 final (Luxembourg: Office for Official Publications of the European Communities).

—— (1991a) 'Communication of the Commission to the Council: the Development and Future of the CAP, Reflections Paper of the Commission,' COM(91)100 final (Luxembourg: Office for Official Publications of the European Communities).

—— (1991b) 'Reform of the Common Agricultural Policy – Legislative Measures to Accompany the Reform of the Market Support Mechanisms,' COM(91)415 final (Luxembourg: Office for Official Publications of the European Communities).

—— (1992) Directorate-General for Agriculture, 'CAP Working Notes 1992: the Reform of the Common Agricultural Policy' (Luxembourg: Office for Official Publications of the European Communities).

—— (1993) Directorate-General for the Environment, *Towards Sustainability. A European Community Programme of Policy and Action in Relation to the Environment and Sustainable Development* (Luxembourg: Office for Official Publications of the European Communities).

—— (1994a) Directorate-General for Economic and Financial Affairs, 'EC Agricultural Policy for the 21st Century,' *European Economy. Reports and Studies*, No. 4.

—— (1994b) 'Growth, Competitiveness, Employment. The Challenges and Ways Forward into the 21st Century,' White Paper (Luxembourg: Office for Official Publications of the European Communities).

—— (1995) Directorate-General for Agriculture, 'CAP Working Notes Special Issues: GATT and European Agriculture,' (Luxembourg: Office for Official Publications of the European Communities).

—— (1996a) 'Progress Report from the Commission. On the Implementation of the European Community Programme of Policy and Action in Relation to the Environment and Sustainable Development. "Towards Sustainability".' COM(95)624 final (Brussels: European Commission), 10 January.

—— (1996b) 'Communication from the Commission to the Council and the European Parliament. European Community Water Policy,' COM(96)58 final (Brussels: European Commission), 21 February.

—— (1996c) 'Communication from the Commission to the Council and the European Parliament. On Trade and Environment,' COM(96)54 final (Brussels: European Commission), 28 February.

—— (1996d) 'Communication to the Council, the European Parliament, the Economic and Social Committee and the Committee of the Regions. The Global

Challenge of International Trade: A Market Access Strategy for Europe,' reproduced in *AE-Europe Documents*, No. 1976/77, 21 February.

COPA/COGECA (1995a) *Agriculture and Forestry: Example of Sustainable Development*
(Brussels: COPA/COGECA).

—— (1995b) *Strategy for the Future Accession of the Associated Countries of Central and
Eastern Europe* (Brussels: COPA/COGECA).

EP (1986) 'Report on Behalf of the Committee on the Environment, Public Health
and Consumer Protection on Agriculture and the Environment,' Rapporteur:
François Roelant du Vivier for the environment committee; Mrs Crawley for the
opinion of the agriculture committee, DOC A2–207/85, EP Working Documents (Strasbourg: EP).

Floyd, William (1995) 'Fiche de Lecture', unpublished memo (Brussels: European
Commission – Cellule de Prospective)

GATT (1994) 'Report on the GATT Symposium on Trade, Environment and
Sustainable Development,' *Trade and the Environment*, TE 008, 28 July.

Grant, Wyn (1995) 'The Limits of Common Agricultural Policy Reform and the
Option of Denationalisation,' *Journal of European Public Policy*, Vol. 2, No. 1,
pp. 1–18.

Haigh, Nigel (1992) *Manual of Environmental Policy: EC and Britain* (London: Institute for European Environmental Policy).

Harvey, David (1994) 'Agricultural Policy Reform after the Uruguay Round,' in K.
A. Ingersent, A. J. Rayner, R. C. Hine (eds), *Agriculture in the Uruguay Round*
(New York: St Martin's Press), pp. 223–259.

Hendriks, Gisela (1994) 'German Agricultural Policy Objectives,' in Rasmus Kjeldahl and Michael Tracy (eds), *Renationalisation of the Common Agricultural Policy?*
(Copenhagen: Institute of Agricultural Economics), pp. 59–74.

Ingersent, K. A., A. J. Rayner and R. C. Hine (1994a), 'The EC Perspective,' in K.
A. Ingersent, A. J. Rayner, R. C. Hine (eds), *Agriculture in the Uruguay Round*
(New York: St. Martin's Press), pp. 55–87.

—— (1994b) 'Agriculture in the Uruguay Round: an Assessment,' in K. A.
Ingersent, A. J. Rayner, R. C. Hine (eds), *Agriculture in the Uruguay Round* (New
York: St. Martin's Press), pp. 260–290.

Josling, Timothy (1994) 'The Reformed CAP and the Industrial World,' *European
Review of Agricultural Economics*, Vol. 21, No. 3, pp. 513–527.

Keeler, John (1995) 'Agricultural Power in the European Community: Explaining
the Fate of CAP and GATT Negotiations,' *Comparative Politics*, Vol. 28, No. 2,
pp. 127–150.

Lenschow, Andrea (1995) 'Policy and Institutional Change in the EC: Environmental Integration in the CAP,' paper presented at the European Community
Studies Association Conference in Charleston, SC, 11–14 May.

—— (1996) *Policy and Institutional Change in the European Community: Variation in
Environmental Policy Integration*, PhD Dissertation, New York University.

Marsh, John S. and Pamela Swanney (1980) *Agriculture and the European Community*
(London: George Allen & Unwin).

Moyer, Wayne H. and Timothy E. Josling (1990) *Agricultural Policy Reform. Politics and Process in the EC and USA* (New York and London: Harvester
Wheatsheaf).

Ockenden, Jonathan and Michael Franklin (1995) *European Agriculture. Making the CAP Fit the Future*, Chatham House Papers (London: Pinter Publishers and the Royal Institute of International Affairs).

OECD (1994) *Agricultural Policy Reform: New Approaches. The Role of Direct Income Payments* (Paris: OECD).

Reeve, Rachel (1994) 'Confronting the Nitrate Problem: The Dutch and British Responses to the 1991 EC Nitrate Directive,' in Michael Wintle and Rachel Reeve (eds), *Rhetoric and Reality in Environmental Policy. The Case of the Netherlands in Comparison with Britain* (Aldershot: Avebury Studies in Green Research), pp. 96–81.

Runge, Ford A. (1993) 'Trade Liberalisation and Environmental Quality in Agriculture,' *International Environmental Affairs*, Vol. 5, No. 2, pp. 95–128.

—— (1994) 'The Environmental Effects of Trade in the Agricultural Sector,' in OECD, *The Environmental Effects of Trade* (Paris: OECD).

Steichen, René (1994) 'The Common Agricultural Policy: The Medium-Term Outlook in the Context of Future Trends in International Trade in Agriculture,' speech given at the Centre for Agricultural Strategy, reproduced in *AE – Europe Document*, No. 1914, 7 December.

Tracy, Michael (1989) *Government and Agriculture in Western Europe 1880–1988*, third edn (New York: New York University Press).

—— (1993) *Food and Agriculture in a Market Economy. An Introduction to Theory, Practice and Policy* (La Hutte: Agricultural Policy Studies).

Urwin, Derek W. (1991) *The Community of Europe. A History of European Integration Since 1945* (London and New York: Longman).

Whitby, Martin (1996) 'The Prospect for Agri-environmental Policies Within a Reformed CAP,' in Martin Whitby (ed.), *The European Environment and CAP Reform. Policies and Prospects for Conservation* (Walingford: CAP International), pp. 227–240.

Wilkinson, Alan (1994) 'Renationalisation: An Evolving Debate,' in Rasmus Kjeldahl and Michael Tracy (eds), *Renationalisation of the Common Agricultural Policy?* (Copenhagen: Institute of Agricultural Economics), pp. 23–32.

WWF (1996) *Article 39 of the Treaty of the European Union. New Objectives for the Common Agricultural Policy of the European Union* (Surrey: WWF-UK).

8

COMPETITIVENESS, ENVIRONMENTAL SUSTAINABILITY AND THE FUTURE OF EUROPEAN UNION DEVELOPMENT COOPERATION

Nick Robins

This chapter explores how Europe's drive for international competitiveness is transforming efforts to improve the environmental performance of European development cooperation programmes. It starts by reviewing the three strategic driving forces for European aid flows to developing countries – national security, economic advantage and the moral imperative of development – within which environmental sustainability has traditionally had a low priority. The chapter then traces the efforts of the European Union to 'green' its aid programme in recent years, before turning to the current crisis facing aid and the competing priorities that could shape its future. Competitiveness is one of these priorities, and the chapter describes three ways in which competitiveness can impact upon the environmental quality of aid programmes. The chapter closes with a set of five conclusions for coping with the EU's triple challenge of competitiveness, environmental sustainability and development cooperation.

Aid, foreign policy and sustainable development

At the June 1992 United Nations Conference for Environment and Development (UNCED – popularly known as the Earth Summit) held in Rio de Janeiro, the governments of the world gave development assistance a central role in implementing the Agenda 21 action plan for sustainable development in the South. Indeed, the UNCED secretariat estimated that $125 billion in the form of concessional financial flows from the North would be needed to turn the words of Rio into action in the developing world. At Rio itself, the European Union (EU) pledged $4 billion for the early implementation of Agenda 21 in developing countries. And although demands from the Group of 77 coalition of developing countries for a substantial 'green fund' were denied by the industrialised world, the Global Environmental Facility, run jointly by the World Bank, the United Nations Development Programme (UNDP) and the United Nations Environment Programme (UNEP) was designated as the mechanism for additional funding of climate change, biodiversity, protection of international waters and ozone layer issues to the tune of $1.2 billion, subsequently increased to $2 billion for 1994–1997.

But even as hugely ambitious goals for increased development aid were being set, the political and economic foundations on which they rested were shifting. Recession hit public expenditure in the North, and aid was a soft target for spending cuts. Economic insecurity also weakened popular commitment for financing 'distant obligations', anyway undermined by a spate of aid finance scandals. With the end of the Cold War and the acceleration of globalisation, a new set of common interests are required to underpin continuing flows of public finance from North to South (ECDPM 1996).

Traditionally, there have been three foreign policy driving forces for aid. The first relates to the promotion of national security, where aid is used as an instrument of foreign policy to project a country's values and defend its interests. It is not by accident that most of Europe's aid agencies are subject to control by their foreign ministries.[1] According to the UN Development Programme, 'so far, the basic motivation for donors to give aid has been to win friends in the Cold War confrontation between socialism and capitalism' (UNDP 1993). The most recent example of this Machiavellian motivation for aid was provided in the aftermath of the Mururoa nuclear testing controversy in 1996, when France boosted financial assistance to the South Pacific to restore political credibility.

Strongly linked to the promotion of national security has been the use of aid to enhance a country's commercial advantage and competitiveness. Over a quarter of all aid is formally tied to the purchases in the donor country. For some EU member states the proportion is much higher, ranging from between 35 and 40 per cent for France and Germany to two-thirds of total ODA for the UK and reaching 80 per cent for Spain. Donors also engage in 'informal tying', so that, for example, Denmark sets itself the goal of a 50 per cent rate of return to Danish business from bilateral aid spending (ACTIONAID 1995).

The third driving force for aid is the moral imperative of 'development', a constantly evolving policy objective, bringing together the goals of poverty relief, human rights, population control and, more recently, environmental sustainability. While development is the stated objective of all aid programmes, the fact that there is still no common foundation for assessing the impact of aid programmes on the relief of poverty or the conservation of natural resources shows that these goals remain somewhat marginal to real aid decision-making: in business language, 'what gets measured gets done', and the qualitative aspects of aid are still not being measured adequately (see ACTIONAID 1996).

There is an environmental dimension to each of these three driving forces. Increasingly, environmental security is being highlighted as one of the core themes for post-Cold War foreign policy. Similarly, the promotion of exports of clean technologies is an important sector within the wider competitiveness agenda, while environmental action is regarded as essential for the long-term relief of poverty in the south. However, in all three, environmental sustainability is only one goal among many, and in terms of the specific links between the competitiveness and sustainability agendas, the key question is whether, in an increasingly competitive and environmentally degraded global economy, the pursuit of Europe's commercial advantage will further erode ties of solidarity between rich and poor and inevitably lead to the undermining of the interests of future generations.

Before this question can be answered, however, it is necessary to examine the mixed results of the EC's attempts to green its aid programme over the past decade.

The greening of European Community aid

Integrating the environment into EC development cooperation

Over the past two decades, the European Community's development cooperation policies and programmes have evolved in a somewhat *ad hoc* fashion to become the world's fifth largest, after Japan, the USA, France and Germany, spending $4.8 billion in 1994; taken together with the member states, the EU provides 45 per cent of total aid from the OECD. In the process, the EC has accounted for a rising proportion of member state aid funds, from just 6.7 per cent in 1970 to 17.4 per cent in 1994; this is likely to rise further in the future (OECD 1996a). For historical reasons, the EC's aid programme is somewhat artificially divided between the seventy African, Caribbean and Pacific (ACP) countries which have signed the Lomé Convention – managed by the Development Directorate (DGVIII) – and Asian, Latin American, Mediterranean nations, and now the Central and Eastern European countries and newly independent states – managed by the External Affairs Directorate (DGI) – through a series of regional strategies and bilateral trade and co-operation agreements. Funding for these agreements comes directly from the EC budget, along with a series of supplementary funds available for all countries and covering food aid, emergency aid, democracy and human rights, co-financing for NGOs, research and development, AIDS, as well as environment and tropical forests. Over the years, aid through the Lomé mechanism has declined compared to other regions. This trend is likely to accelerate after 2000, as the importance of Eastern Europe, the Mediterranean and Asia grows (see Tables 8.1 and 8.2).

The importance of the environmental dimension of Community development cooperation has grown in line with the growing financial significance of EC aid (Agrasot 1995, Robins 1996a). In part, this has been a response to concern over poor environmental performance. A recent survey by the European Environmental Bureau, for example, found that 'for a good number of NGOs in the South, European cooperation appears like a "secret monster" which manifests itself by "money munching" projects far away from the concerns of beneficiaries' (EEB 1993). The Court of Auditors has sometimes criticised individual projects, for example, the lack of environmental appraisal for a significant road building scheme in Cameroon (CoA 1992). Parliamentarians along with a number of environment and development organisations have also highlighted environmental and social shortcomings of controversial initiatives such as the Carajas steel programme in Brazil, the Kibale forest scheme in Uganda and the Risonpalm palm oil plantation project in Nigeria (Hayter 1988, Worthington 1993, CLO 1994, WDM 1994).

Prompted by these concerns and its own internal process of policy evolution, there has been a gradual upgrading of the EC's environmental policy goals, reflected in successive Environmental Action Plans and Development Council Resolutions. Since 1987, the Commission has aimed to make 'an environmental reflex' a natural part of mainstream aid administration, although an independent report two years later found that some staff still viewed the environment as 'just another burden on top of an already heavy work load' (CEC 1987, Wenning 1989). At the height of the wave of environmental concern in the late 1980s and early 1990s, the Development

Table 8.1 European Community aid by region: 1990–1994

Region	1985–1989	1990–1994	OECD average (1993)
Sub-Saharan Africa	64 per cent	55.7 per cent	33 per cent
South and East Asia	14 per cent	10.1 per cent	31 per cent
Oceania	3.2 per cent	1.8 per cent	–
North Africa and Middle East	7.2 per cent	14.2 per cent	–
America	9.2 per cent	11.4 per cent	12 per cent
Europe	2.3 per cent	7.05 per cent	–

Table 8.2 European Community aid by region: 1995–2000

Region	Billion ECUs
Africa, Caribbean and Pacific	14.6
Asia and Latin America	4.0 (estimate)
Mediterranean	5.26
Central and Eastern Europe	6.7
Former Soviet Union	2.7

Council passed more resolutions on the environment than on any other issue, deciding in 1990 that environmental protection was one of the 'priorities of development assistance', coinciding with the heightened importance given to international environmental action at the June 1990 Dublin Summit.

The Lomé Convention and the environment

The EC's Lomé Convention framework for seventy African, Caribbean and Pacific (ACP) countries remains the central pillar of Community development cooperation. Managed outside of the normal EC budgetary framework, Lomé provides a package of trade preferences and development assistance through successive European Development Funds and loan agreements with the European Investment Bank. The fourth Lomé Convention signed in December 1989 allocated ECU 12 billion for 1990–1995, of which the Commission estimates that approximately 10 per cent will be allocated to projects with a primary environmental objective. After much haggling, funding of ECU 14.635 billion was agreed for 1995–2000 at the Cannes European Summit in June 1995. About 65 per cent will be allocated to development projects on a country by country basis.

The agreement of the fourth Lomé Convention in 1989 marked a step change in approach towards environment and sustainable development. Following the Brundtland Report and the inclusion of environmental provisions of the Single European Act, the new Convention incorporated an overall commitment to a 'sustainable balance between economic objectives, rational management of the environment and the enhancement of human and natural resources', along with a series of specific environmental provisions laying out the principles and procedures for integrating the environment into Lomé aid activities.

The post-Rio emphasis on raising the environmental quality of aid has coincided with the imperative of improving the overall effectiveness of Lomé funded assistance. Thus, in 1993 a new Project Cycle Management (PCM) approach was introduced as a way of bringing greater rigour to aid administration, for example, by ensuring that projects are really targeted at the needs of intended beneficiaries and that all the factors for the sustainability of a project (including the environment) are taken into account. Based on the PCM, annual monitoring forms have to be completed for each project, including information on environmental sustainability. Using the PCM approach, the Commission issued an Environment Manual, also in 1993, to guide decision-makers in the Commission and ACP partner countries on how to integrate environmental factors at each stage of the project cycle from conception to final evaluation (CEC 1996a, 1996b).

Efforts have also been made to ensure that environmental considerations are taken on board at a more strategic level during the preparation of the five-year national indicative programmes which lay out the priorities for each ACP country. According to the Commission's review of the Fifth Environmental Action Programme, analysis of the national and regional indicative programmes for 1990–1995 which account for the bulk of Lomé aid financing shows that the goal of environmental integration was only partially achieved (CEC 1994b). Only a few included sustainable development as a cross-sectoral issue at the strategic level, although almost all rural development programmes incorporated environmental factors (CEC 1996a). To tackle this, forty Environmental Synopses of ACP countries have been circulated to assist in country level planning, and Environmental Guidelines were prepared to guide the programming for 1995–2000. These incorporate the lessons of Rio for aid management, stressing the importance of both integrating the environment into the design and implementation of all Community activities, and integrating social and economic factors with initiatives aimed specifically at environmental protection. The guidelines set the core goal of strengthening the in-country ACP human, technological and institutional environmental capacities in the public, private and civil sectors (NGOs).

Asia and Latin America

The Community has taken the same twin-track approach to improving the environmental performance of its cooperation with Asian, Latin American and Mediterranean countries: increasing the proportion of resources targeted at meeting environmental needs and improving the procedures for environmental assessment. At least 10 per cent of technical and financial assistance resources for Asia and Latin America between 1992 and 1996 will be aimed at projects and programmes with a primary environmental objective. Internal procedures for environmental impact assessment (EIA) were established in November 1991, and then amplified in the guide for project management in July 1993. All projects over ECU 1 million are now routinely screened for environmental implications; full EIAs are then carried out for those projects that pose potential problems, amounting to some 16 per cent of the total in 1994. Projects without proper environmental screening are now refused financing, and then have to be resubmitted for approval once they integrate the recommendations of the EIA as environmental management plans. In general, the Commission believes that there is a clear trend towards projects with positive environmental impacts or projects which have had preliminary EIA in the preparation phase (CEC 1996a, 1996b).

In some senses, the process of mainstreaming the environment into EC development activities appears to have gone further and faster in the case of Asia and Latin America. For example, the Commission's review of the Fifth Environmental Action programme argues that

> the process of monitoring EIAs for aid to Asia and Latin America needs to be made universal for all the Community's aid, so that it is easy to demonstrate that EIAs have a real impact on the scope and content of projects.
>
> (CEC 1996a)

One reason for this could lie in the greater involvement of the EC's Environment Directorate in the non-ACP cooperation agreements. Thus, as part of European cooperation with ASEAN (Brunei, Indonesia, Malaysia, the Philippines, Singapore and Thailand), a specialised forest and environment working group has been established to develop a joint response to the problems of the loss of tropical forests in the region.

The mounting importance of China and India in global environmental negotiations, particularly for climate change, due to their ambitious coal-fuelled growth plans, has meant that the environment has become an important strand of overall EC relations. In China, environmental cooperation was one of the four priority areas identified by the Council of Ministers in a December 1995 resolution on China–Europe relations, following Commission proposals for a long-term strategy (CEC 1995a; the pivotal role of Asian environmental issues is also discussed in Chapter 5). A joint environmental working group has been set up by the Commission's Environmental Directorate (DGXI) with its Chinese counterparts, which has carried out a fact finding mission to draw up priorities for future cooperation, focusing on integrated environmental management in the industrial sector. Sub-Saharan Africa, by contrast, has far less significance for EC international environmental policy, with the exception of tropical forest management issues in Central Africa and to a lesser extent, widespread soil degradation and desertification. Somewhat surprisingly given the long-established nature of the Lomé framework and its commitment since 1989 to sustainable development, there is no regular forum for information exchange and priority setting on environmental issues: the last major initiative appears to have been the Euro-African Environment and Development ministerial conference held in 1988.

Development cooperation and international environmental agreements

In the run-up to Rio and beyond, EU donor agencies have been exploring ways of supplementing the traditional approach of integrating the environment into their development programmes, with more targeted 'environmental assistance'. Thus, as part of the restructuring of the Global Environmental Facility, France has established its own Fonds Français pour l'Environnement Mondial (FFEM), with an allocation of an additional Ffr 440 million for 1993–1997. The FFEM works in the same four priority areas as the GEF – biodiversity, climate, international waters and ozone depletion – and aims to catalyse existing French aid to do more on these issues. The FFEM supplements mainstream French aid for the environment, which has another three priorities, primarily for Africa – maintaining marine fisheries; conserving tropical forests; and supporting natural parks – which sit alongside a number of smaller scale programmes for environmental policy support, freshwater, the urban environment, renewable energy, desertification and cooperation with NGOs, which together amounted to Ffr 230 million in 1994 (RF 1995).

In Denmark, a Danish Fund for Environment and Development (DANCED) was established after Rio to help tackle global environmental problems and implement the UNCED agreements (DANIDA 1994). This follows Denmark's decision to allocate an extra 0.5 per cent of national income by 2002 for environmental action and disaster relief. DANCED focuses on seven key themes – urban areas, sustainable energy, agriculture, water resources, forests, biological diversity and coastal zones – in countries not eligible for environmental support under normal Danish development assistance. In all, developing countries will receive 25 per cent of the fund, a further 25 per cent for Central and Eastern Europe, and the remaining half for the Arctic region. Initial efforts have been concentrated in South-East Asia, in particular Malaysia and Thailand, and Southern Africa. Importantly, DANCED is managed by the Ministry of Environment and Energy in close cooperation with the DANIDA development agency. Funding in 1995 totalled Dkr 295 million, rising to Dkr 500 million in 1997 (DANCED 1996).

A set of three 'Sustainable Development Agreements' signed by the Netherlands in 1994 with Benin, Bhutan and Costa Rica are perhaps the most innovative response to UNCED. The ten-year agreements are based on the principles of reciprocity, equality and participation, and in a departure from traditional development relations, the agreements focus on the changes needed in the North, as well as ways in which aid can support progress towards sustainability in the South. The agreements are managed by independent foundations in each country, with a total budget of Dfl 30 million in 1995 (Ecooperation 1996). These deliberately experimental agreements parallel more traditional Dutch support for environmental improvement in the South, accounting for Dfl 335 in 1995, just under a quarter of total aid flows (T&B Consult 1996).

By contrast, neither Britain nor Germany has established new funding mechanisms for implementing the UNCED agreements. In the UK, special attention has focused on five priority areas – forestry, biodiversity, population, sustainable agriculture and energy efficiency – with a special emphasis given to cross-cutting themes of participation in the aid process (Chalker 1994). Germany has no special funding earmarked for environmental co-operation with developing countries, although approximately a quarter of its financial and technical assistance is spent in the field of environment and natural resources.

At the EC level, resources for environmental protection have also been rising as a proportion of overall spending to about 10 per cent of the total; in 1994, the Commission estimated that over ECU 1 billion of Community funds were spent outside the EU on environmental projects (see Table 8.3). Within this, tropical forests are the central environmental issue for the Community's external environmental assistance efforts, bringing together the commitments under Agenda 21, the Rio Declaration and Forest Statement, along with the Climate, Biodiversity and Desertification Conventions. In the past five years, three interlinked initiatives have been taken to push ahead with tropical forest conservation by the Community. A new Tropical Forest Regulation was adopted in December 1995, guaranteeing ECU 50 million a year up to 2000. Key themes for funding will include the conservation of primary tropical forests, sustainable management of production forests, the definition and development of forest certification systems and support for indigenous peoples. A Tropical Forest Protocol was also added to the Lomé Convention in November 1995, concentrating on boosting the timber trade from sustainably managed forests, supported by EDF finances. Finally, special incentive arrangements have been established under the new General System of Preferences trade regime for non-Lomé states to reward those developing countries that adopt and apply sustainable forest management practices.

Assisting sustainable development?

For the past decade, the Commission, along with other European development agencies, has been struggling to implement a 'first generation' set of environmental reforms to its aid programmes. This has involved integrating the environment into mainstream Community funding through changes to project management and procedures change, a theme underscored by the Fifth Environmental Action Programme (CEC 1993a), and investing a growing proportion of overall aid into particular environmental priority areas, notably desertification and deforestation. Here, some possible success

Table 8.3 European Community aid spending on the environment 1994

Region	Million ECU
Africa, Caribbean and Pacific	536
Mediterranean	26
Asia	105
Latin America	28 (estimate)
Ecology and Forest Budget Lines	70
Central and Eastern Europe (CEECs)	75
New Independent States (NIS)	12
Nuclear Safety (CEECs/NIS)	116
Research and Development	45 (estimate)
LIFE Fund – Baltic and Mediterranean	5
TOTAL	1018

stories are now emerging from Community funding, such as the PFIE (Programme Formation et Information sur Environnement) environmental education programme for schoolchildren in the Sahel and the ECOFAC regional network of forest conservation and protected areas in Central Africa.

However, progress with this 'first generation' agenda has been held back by three major constraints: a lack of public accountability, insufficient environmental capacity, and weak coordination with the aid programmes of member states. Although the Court of Auditors screens the financial efficiency of European spending programmes, there is still no public accountability for the environmental performance of EC aid spending. The Commission has so far ignored the European Parliament's request for an annual report on how the environmental provisions of the Lomé Convention have been implemented (ACP-EC 1994, Ewing 1996). The European Union thus lags behind the World Bank, which has now published five annual environmental reports; equally, the EC has yet to face an equivalent to the sustained environmental critique over the past decade which has forced environmental issues up the Bank's agenda (Rich 1994). Public accountability provides an important feedback mechanism for improving the quality of public spending, and without it, the quality of aid is left up to the professionalism of aid officials. And here lies the second obstacle to the greening of European aid: the continuing lack of capacity among EC staff to implement its environmental policies. A recent OECD review of Community aid found that 'the Commission continues to suffer from a lack of environmental specialists and is frequently not in a position to carry out the necessary environmental impact assessments of the development projects it supports' (OECD 1996b). Financial pressures to contain staff levels in the Commission mean that this capacity gap is unlikely to be filled in the near future.

The Commission's capacity problems are exacerbated by a continuing inability to define in clear terms the 'value added' of European aid compared with that provided by the member states: the EC is seen as a sixteenth donor, additional to the member states bilateral programmes. The Treaty of European Union, which included for the first time a clear constitutional basis for aid, sought to rectify this by seeking to boost the effectiveness of European aid through better coordination with member states and improved coherence with other policy areas, most notably the Common Agricultural Policy. Some progress has been made in drawing up common policy statements on poverty, gender and health and coordinating operational activities in six pilot developing countries. But the continuing desire among member states to use aid to pursue national security interests abroad and promote national champions in export markets means that the scope of coordination will remain limited.

Much still needs to be done, and in its most recent environmental resolution in May 1996, EU development ministers called on the Commission to step up the 'tempo and momentum of its efforts towards upgrading its EIA procedural and implementing capacities' and underlined the need for 'the systematic application of EIA procedures before the formal adoption of all Community development projects'. But even while this 'first generation' agenda remains unfinished, a new 'second generation' agenda is emerging, giving greater importance to competitiveness concerns in the shaping of the environmental dimension to European aid.

Competing priorities for development assistance

Development cooperation policy and practice is going through an uneasy period of reassessment as both North and South look hard at the fundamental reasons for aid. Ashok Khosla, director of Development Alternatives in India, has estimated that over the past five decades more than a trillion dollars has been spent in aid, but states that 'it is difficult to avoid the conclusion that this world-wide collaboration has been a failure. Many today would say it was an outright disaster' (Khosla 1995). For Khosla and others seeking to make a reality of sustainable development, the task is to make aid smaller, more decentralised and participatory, sensitive to issues of gender, social justice, environment and culture. Central to this current battle of ideas and interests are the issues not just of the new balance to be struck between development, security and competitiveness objectives, but whether these goals can coexist without undermining prospects for environmental sustainability.

The starting point for the rethink is the fall in overall aid flows, which have often been dramatic, as in the United States, where the Republican dominated Congress cut official development assistance by almost a fifth since 1992. But in the European Union as well, aid levels in most member states – Belgium, Finland, Germany, Italy, the Netherlands, Spain, Sweden and the United Kingdom – were lower in 1994 than in 1992. Spending by the European Community itself has risen slightly, largely because funding levels for the 1990s had been agreed at the December 1992 Edinburgh Summit. This general climate of financial stringency, combined with post-Maastricht manoeuvring over questions of sovereignty and the control of aid, spelt doom for EU's $4 billion UNCED aid pledge, with the European Commission admitting at the end of 1994 that the Union had failed to 'reach an agreement on a common and coordinated framework for the implementation of the financial commitment made in Rio' (CEC 1994b). Overall, OECD aid flows to the South are now at their lowest levels for two decades, and they are unlikely to rise in the near future. Maastricht convergence criteria for a single currency mean that for the EU in particular, aid volumes will remain stagnant into the next decade. Thus, rather than marking a new North–South grand bargain on how to manage the Earth's limited carrying capacity in a wise, fair and prudent fashion, the Earth Summit was 'the death knell of international welfarism' (Sandbrook 1995).

Over the past five years, a series of international summits – Rio (environment), Cairo (population), Copenhagen (social justice), Beijing (gender), Istanbul (cities) and Rome (food) – have attempted to reignite the development rationale for aid. The recent Caring for the Future report has also reasserted the moral basis for aid, stating that 'in future, grant aid and conditional loans should be more consciously targeted towards sustainable improvements in the quality of life – especially by those currently not enjoying it, the poorest strata of the population in low income countries' (ICPQL 1996). From an environmental perspective, this focus on meeting needs has also been rising in importance, largely thanks to the use by Friends of the Earth and others of the concept of 'environmental space', which states that each person on the planet has a right to an equal share of the Earth's carrying capacity (Milieudefensie 1992, FOEE 1995). In its report on the implications of Rio for aid policy, the Dutch Government's national development advisory council concluded that priority should henceforth be given to ensuring an equal share of this eco-space to the South and reorienting development programmes accordingly (NAR 1993).

The security aspects of foreign aid, having receded after their prominence during the years of the Cold War, are again asserting themselves. Louk Box, director of the European Centre for Development Cooperation Management, thus argues that 'security and no longer development will become the primary consideration of cooperation. The types of cooperation which promote "common human security" between North and South will gain in importance' (Box 1995). As a trading bloc first and foremost, security questions for the European Union have traditionally focused on how to deal with Europe's often acute dependence on external sources of energy and raw materials. Born at the height of Third World commodity power in the 1970s, the extensive trade and aid provisions of the Lomé Convention no longer respond to Europe's strategic security interests, which have been replaced by concerns about the safety of its 'near abroad' in Eastern Europe and the Mediterranean. Here, environmental factors are rising in importance as a possible new determinant for aid, with pre-emptive efforts to achieve long-term sustainable development being recommended as a way of defusing the security implications of environmental degradation and subsequent migrations: 'export the wherewithal for sustainable development for communities at risk or import growing numbers of environmental refugees' (Myers 1995). But the environmental security argument is double-edged, with environmental degradation sometimes being used to justify enhanced military forces rather than better long-term aid (Robins 1996b).

But it is economic globalisation that is perhaps the strongest pressure for rethinking aid, so that it becomes more fully supportive of a donor's competitiveness. There is growing concern that this could 'result in aid programmes reflecting domestic export interests over the sustainable development needs of recipient countries' (ACTIONAID 1995). At the EU level, few links were made by the landmark Delors White Paper on Growth, Competitiveness and Employment, between the promotion of Europe's commercial advantage and its relations with the South, particularly in terms of the implications for EU development cooperation (CEC 1993b). The global vision presented was of a Triad of dominant economic powers (EU, Japan and the USA), with the rising challenge of the Asian emerging economies and Eastern Europe alongside. The poor South, particularly Africa, was not picked up by the Commission's competitiveness 'radar'. This has two implications: first, it makes the relatively economically-unattractive ACP countries vulnerable to a potential switch in aid resources to Asia and Latin America, and second, it is likely to lead to a shift in aid resources away from poverty-focused development in favour of growth-oriented economic cooperation through trade, technology transfer and investment promotion. This was particularly marked in the EU's recent strategy for Asia, and was further reinforced at the April 1996 Asia–Europe Summit in Bangkok, attended by twelve EU premiers, when the focus of attention was on 'a new comprehensive Asia–Europe Partnership for Greater Growth', stressing trade, investment and business links: cooperation for development and environment were relegated to a subsidiary place (CEC 1994a).

There are a range of links between this new competitiveness agenda and environmental sustainability. Although the term 'competitiveness' rarely appeared – if at all – in the UNCED agreements, the issue of how to resolve competing economic claims to the global environment was central to the often heated negotiations on tropical forests, intellectual property rights for biotechnology and the transfer of clean technologies. Since Rio, these issues have come out into the open, particularly in the area of changing consumption patterns and trade, where developing countries fear that efforts by the North to shift to more efficient and less polluting growth models will both reduce demand for their commodity exports and impose 'green protectionist' barriers to their manufactured goods. The EC, along with a number of EU member states, are also explicitly seeking to search out 'win-win' solutions in the promotion of environmental technologies and techniques through a range of 'second generation' aid initiatives. The Commission, for example, now openly acknowledges the importance of 'examining the role of cooperation in relation to clean and efficient technologies, which not only bring solutions to problems in developing countries but also provide European producers with investment opportunities' (CEC 1996b). Critically, the accent given to competitiveness concerns tends to focus environmental action at the clean-up or prevention of the 'effluents of affluence': in other words, greening the growth trajectories of rapidly expanding emerging economies (see Chapter 5). This could divert attention from the core concern of sustainable development agenda, which is the environmental needs of those living in poverty.

The next section explores whether this ambition for 'win-win' benefits is being realised in practice, and its implications for developing countries.

Competitiveness and sustainable development: who wins, who loses?

Competitiveness and development assistance

To an extent, the increased emphasis on commercial factors is nothing new. Rather, it is an intensification of the longstanding use of development assistance funds and other policy levers for the external support of the donor's own industries and economic interests. This takes place at two main levels. At the micro-level, competitiveness is promoted by European governments through the allocation of export credits, loans and the formal or informal tying of aid to the purchase of national goods and services. Generally, the multilateral approach adopted by Community aid funds has been favoured by developing countries, as they are much less tied to national purchases. But although the EC does not publish any equivalent targets, it states that 'for every ECU 100 spent on aid, the Community recovers ECU 48 in the form of projects, supplies and technical assistance purchased from European companies' (CEC 1996c). In the specific case of the Lomé Convention, ACP companies won only 20 per cent of contracts concluded under the Sixth European Development Fund (EDF) and 17.6 per cent under the Seventh EDF, with the remainder going to European Union companies (CEC 1995b).

Tied aid has a poor reputation. The lack of competition can lead to the over-pricing of goods and services by about 15–30 per cent. While individual firms certainly profit from such government subsidies, the evidence of benefit for either donor or recipient is often less clear. In the case of the UK's Aid and Trade Provision, a government evaluation found that 'very few real economic benefits' were generated (ODA 1991). In the case of EC aid, research in the Philippines has criticised the lack of accountability and transparency of some Community programmes, and an over-reliance on European consultancies rather than local agencies (Santos 1995, Putzel 1996). Senior officials in both donor agencies and recipient states stress that five decades of misguided technical assistance have seriously harmed recipient countries (Jaycox 1993). In the EU context, some fierce criticisms are emerging, particularly of the technical assistance provisions contained in the Lomé Convention framework for relations with Africa, the Caribbean and the Pacific (Bossuyt 1995). In an attempt to prevent the worst side-effects of aid tying, the OECD has agreed the Helsinki Principles, a 'gentleman's agreement' to establish a level playing field among donors. But in the rush to conclude contracts – often in competition with other donor countries – governments can skimp on basic economic, social and environmental assessments, leading to the 'Pergau Syndrome' (see p. 205).

At the macro-level, European governments promote the competitiveness of their industries and economies through a variety of sectoral and structural policies that aim to give comprehensive advantages over developing country producers. The OECD has bemoaned the way in which development objectives are undercut by 'inconsistent policies in trade, investment, environment, debt, arms sales, agriculture and other areas', concluding somewhat sadly that these inconsistencies are due to the fact that development is seen as 'a separate and distinct field of secondary importance' (OECD 1996b). For the EU, the classic case of such policy incoherence is the agricultural sector, where domestic subsidies and export promotion have had severe repercussions on the economic prospects of developing countries, undermining the EU's own development efforts (the CAP's environmental damage within the EU is discussed at length in Chapter 7). As part of the new provisions of the Maastricht Treaty, the Community is obliged to take account of its development cooperation goals in other policy areas to ensure policy coherence. Following NGO reports of the damage done by the dumping of EC beef in West African markets, External Affairs Commissioner Marin was forced to admit that 'European exports seriously hurt local production, the regional trade and livestock development projects financed by the EDF' (OECD 1996a). Indeed, the EU still appears to be spending more on subsidising exports than on aid to agriculture in sub-Saharan Africa, negating the aid it is providing (Woodward and Pryke 1995). According to a recent report from SOLAGRAL, this has led to 'an unsustainable contradiction' testing the very credibility of the Union (SOLAGRAL 1996).

Competitiveness, development and the environment

There appear to be three main ways in which the drive for competitiveness in development can impact upon efforts to achieve environmental sustainability.

Where the drive for competitiveness undermines the environment

There is a clear risk that the pursuit of commercial advantage by governments in their aid programmes could compromise other objectives, including environmental sustainability. The UN's Food and Agriculture Organisation has concluded, for example, that the provision of toxic pesticides as part of donor aid programmes has left Africa with a $100 million bill for clearing up obsolete and dangerous stockpiles (see the discussion of EU export policy in Chapter 3). An in-depth study by the US Office of Technology Assessment – recently abolished by the Republican Congress – into the growing emphasis on the transfer of environmental technologies using aid funding, warned that 'export promotion goals could overshadow environmental and developmental goals unless adequate safeguards are in place' (OTA 1993). For example, 'over-emphasis on export promotion could bias projects toward overly expensive infrastructure, with more sophisticated technology than needed to meet basic human needs' (OTA 1993). The OTA recommended that donors should put in place four main safeguards:

- Carry out environmental studies to identify real needs and not just commercial opportunities;
- Apply 'first generation' environmental assessment procedures to export promotion activities;
- Undertake independent evaluations of the environmental performance of technologies;
- Make provisions for the operation and maintenance costs of technology transfer.

In the case of the EC, environmental assessment considerations have been fully integrated in the European Community Investment Partners (ECIP) programme, which aims to create viable joint ventures between EU and Asian and Latin American companies. Since 1992, all of the more than 600 ECIP joint venture proposals have been screened for environmental impacts. This has resulted in full environmental assessments for 14 per cent of the projects presented for financing. However, more generally, there appears to be considerable 'hesitancy' by OECD governments to applying environmental measures to export promotion activities, with fears expressed that this may 'excessively complicate the efforts of exporters . . . [and] may discourage firms from exporting or delay their entry into the international market, resulting in lost sales' (OECD 1994). Most OECD export credit agencies are simply unable to comply with the Montreal Protocol trade restrictions on exports of technologies for producing or using ozone-depleting substances, due to lack of expertise.

Without stringent analysis and public accountability, public aid can be misused to support economically unviable and environmentally damaging projects. The scandal of the Pergau hydro-electric scheme in Malaysia, funded to the tune of £280 million by Britain's aid and export credit programmes, highlights the risks. Although the project was judged an abuse of aid funds by officials and its environmental impacts had not been adequately assessed, political and commercial imperatives linked to arms sales and the interests of British construction companies seeking to build the dam were overwhelming. However, in a landmark case, the UK Government's decision to grant aid to the project was found to be unlawful and payments from the aid budget halted. As competitive pressures intensify, so the risks of a spread of the 'Pergau Syndrome' will mount. The European Union will need to look closely at whether action is needed to ensure that member state export promotion activities comply with the Maastricht Treaty's provisions on the environment, and so establish a level playing field for all.

Where competitiveness and environment can work together

Most industrialised countries are looking at ways of generating 'win-win' benefits for their own industry and developing country environments through the aid-assisted transfer of clean technologies and environmental expertise. One of the three objectives of French environmental assistance is to promote 'le savoir faire français en matière d'environnement', for example, through support for satellite tracking technologies (RF 1995). The Danish DANCED fund also aims to 'act as a catalyst for the commercial utilisation of the potential of Danish companies', as well as respond to local needs and priorities (DANCED 1996). DANCED will thus be used, for example, to apply Danish expertise in cleaner technology through the establishment of a Clean Technology Information Centre in Thailand, focusing on environmental auditing in small companies. The joint EC–Singapore Regional Institute for Environmental Technology (RIET) follows a similar model, having a twin goal to 'promote effective environmental technology throughout Asia and to facilitate mutually beneficial economic cooperation' (RIET 1996). RIET sponsors environmental management training programmes and finances studies on environmental policy questions such as Thailand's ecolabelling needs. It also acts as a marketing vehicle for European environmental technology companies. Eventually, the aim is for RIET to be self-financing.

In these and other cases, it is still too early to assess the extent of mutual benefits gained by such programmes. But a degree of caution is required to ensure that export interests do not overwhelm local needs. There is rising concern in particular at the export of clean-up technologies rather than techniques to conserve resources and prevent pollution, a culmination of a three-stage process of exporting toxic industrial development from North to South:

- first, economic 'development' is exported through free trade policies and financing by multilateral and bilateral agencies;
- second, environmental regulations to control the excesses of this development are introduced;
- finally, 'environmental' technology and services are exported to service these regulations.

(Karliner 1994)

A particular priority, therefore, is for environmental agencies to ensure that an emphasis is placed on clean technology in publicly-funded environmental assistance programmes, rather than encouraging the export of potentially inappropriate end of pipe equipment.

Where aid can help the South's environmental competitiveness

Increasingly stringent environmental standards and preferences in developed countries can act as competitive barriers to developing country producers (see Chapter 6). Much can be done by environmental authorities not only to improve the transparency and accessibility of their activities, but also to respond to developing country advantages. For example, the Federal Environmental Agency in Germany is now developing criteria for awarding the Blue Angel ecolabel to products made from rattan and jute in developing countries. Development agencies can also help to overcome these barriers by actively promoting the environmental upgrading of developing country exports. For example, the EC's Tropical Forest Regulation and the Lomé Tropical Timber Protocol both stress the importance of supporting timber certification schemes as a way of enabling producers of wood products from sustainably managed forests to gain market share and premiums in Europe's environmentally conscious markets (Dieterle 1996). Other donors have been working to promote more environmentally-sound cotton production projects. The Dutch Sustainable Development Agreement with Benin, for example, is examining the possibility of encouraging a switch to better environmental practice in return for the Netherlands agreeing to purchase a guaranteed amount at a guaranteed price. Swedecorp, the former commercial branch of the Swedish aid agency, has designed a programme to ensure that developing country cotton producers do not lose out as environmental preferences rise in Europe, which included funding an organic cotton pilot project in Uganda, finding a Swedish buyer for the output and providing consultancy support for certification and labelling (IIED 1995).

Conclusions: can cooperation and competition be reconciled?

Five years after Rio, the context for assisting progress towards sustainable development in the South has changed markedly. Old touchstones inherited from colonial obligations and the geopolitical imperatives of the Cold War have now dissolved. A new rationale for aid is required, which can help to focus attention on the needs of developing countries, and establish how these can be supported by external financial assistance from the industrialised world in a spirit of mutual self-interest. At a European level, the Union is currently struggling to develop a stronger Common Foreign and Security Policy stance as part of the Maastricht Treaty. This offers an important opportunity for clarifying the balance between self-interest and cooperation in Europe's external strategy. In addition, the Union has also launched a major review of the Lomé Convention to prepare the ground for its replacement in 2000, and will soon have to start negotiations on the successor to the Delors II budget agreement, again providing critical openings for restating the purposes of aid.

What the European Union now needs is a serious public debate about the balance to be struck between security, competitiveness and development in its cooperation activities for the next century. Such a debate would also need to encompass how these competing pressures for aid enhance or degrade the environment, and how they offer greater access to natural resources for the world's poor. New thinking is just starting to emerge on the tensions and dilemmas faced in trying to reconcile these objectives. The EU as a whole needs to address these issues to avoid making potentially costly overlaps between Community and member state aid programmes and ensure a 'level playing' field across the Union. Five themes have emerged from this chapter, which could be used as a starting point for such a wider reflection.

1 Commercial interests have long been an important driving force behind development cooperation, and these are set to intensify under the new competitiveness banner. There is so far no direct evidence that the drive to enhance Europe's competitiveness has led to a reduction in European Union aid volumes, although in the general push to reduce public spending, development has been a 'soft' target for cuts. In terms of the regional balance of aid, competitiveness concerns do seem to be driving aid away from Africa, the Caribbean and the Pacific, in favour of South-East Asia, Latin America, the Mediterranean and Central and Eastern Europe. Cooperation focused on promoting development is also giving

way to economic cooperation focused around promoting trade and investment.

2   Environmental sustainability has achieved a high political profile in European development policy over the past decade, and efforts have been made to integrate the environment and increase aid allocated to environmental projects. However, progress has been constrained by a lack of public accountability for the environmental performance of EC aid, insufficient skills and staff to implement the policy, and weak coordination with the member states. In the future, greater emphasis will be placed on seeking 'win-win' solutions that help to promote exports of European clean technologies and on linking environmental degradation to European security concerns. By contrast, at a time when the status of aid is at historically low levels, the case for aid to reduce poverty and promote long-term sustainable development needs restating.

3   Little policy thought has yet been given to this competitiveness, environment and development nexus. Three main linkages between competitiveness and environment can be observed in European development cooperation:

- there is the negative 'Pergau Syndrome', necessitating mandatory preventative environmental action for all development cooperation funding, but especially finance geared towards export promotion;
- there is the mutually beneficial potential of 'environmental assistance', requiring honesty between donors and recipients if local needs and export interests are to be balanced;
- there is the currently small scale 'altruism' of donors boosting the environmental competitiveness of southern producers.

4   Top priority still needs to be given to guaranteeing rigorous environmental appraisal and management for all aid projects, reinforced by public accountability for environmental performance through regular reporting. Given the growing capacity constraints within European development agencies, the role of environmental ministries, agencies and at the EC level, DGXI, should be reinforced in three main areas:

- first, strengthening the environmental dimension of regional and country aid strategies;
- second, providing technology evaluation services for export promotion programmes;
- third, managing specialised 'environmental assistance' funds, additional to existing aid funds and targeted at raising the environmental capacities of key export sectors in developing countries affected by new environmental regulations in the EU (e.g. ecolabelling, packaging) or changing consumer preferences.

5   Action to coordinate EC and member state environment and development cooperation activities in the spirit of Maastricht also needs to be
bolstered, particularly to ensure that all export promotion activities are
subject to environmental screening, and that environmental assistance
measures do not breach state aid legislation. A more forceful European
approach to financing the implementation of international environmental agreements in the South is required, particularly for climate
change and desertification.

Taken together, these measures could help to ensure that while aid will
never be free of the drive for competitiveness, at least it can be steered in
such a way that environment and development goals are met.

Notes

1   The European Community (now the EU) is a notable exception, having had an
aid programme since its inception in the 1950s, but only developing a formal
foreign policy capacity in the 1990s with the Maastricht Treaty.

Bibliography

ACP–EC (1994) Resolution on Sustainable Development, Joint ACP–EC Assembly,
Strasbourg, 14–18 February.
ACTIONAID (1995) *The Reality of Aid* (London: Earthscan).
—— (1996) *The Reality of Aid* (London: Earthscan).
Agrosot, Paloma (1995) *La Mise en Oeuvre des Accords de Rio et la Communauté
Européenne* (Brussels: European Environmental Bureau), January.
Bossuyt, Jean (1995) 'The Future of EU–Africa Development Cooperation, ECDPM
Working Paper 95–2, European Centre for Development Cooperation Manage-
ment, Maastricht.
Box, Louk (1995) 'De L'assistance à la Coopération,' *Courrier de la Planète*, No. 33,
Solagral, Paris.
CEC (1987) 'Prise en Compte de la Dimension Environnement dans la Politique
Communitaire de Développment,' SEC(87)1838 (Brussels: Commission of the
European Communities), 19 October.
—— (1993a) *Towards Sustainability* (Brussels: Commission of the European
Communities).
—— (1993b) *White Paper on Growth, Competitiveness and Employment* (Brussels:
Commission of the European Communities).
—— (1994a) *Towards a New Asia Strategy*, COM(94)314 final (Brussels: Commis-
sion of the European Communities), 13 July.
—— (1994b) *Interim Review of the Fifth Environmental Action Programme*,
COM(94)453 final (Brussels: Commission of the European Communities), 30
November.
—— (1995a) *A Long-Term Policy for China–Europe Relations*, COM(95)279 final
(Brussels: Commission of the European Communities), 5 July.

—— (1995b) *Financial Cooperation under the Lomé Conventions – Review of Aid at the End of 1994*, Development Information No. 81 (Brussels: Commission of the European Communities), August.

—— (1996a) *Progress Report on the Fifth Environmental Action Programme Review* (Brussels: Commission of the European Communities).

—— (1996b) *Report from the European Commission to the UN CSD* (Brussels: Commission of the European Communities), April.

—— (1996c) *20 Questions and Answers – The Lomé Convention*, Development Information No. 84 (Brussels: Commission of the European Communities), August.

Chalker, Linda (1994) 'Turning Rio into Reality: Two Years on from the Earth Summit,' Speech at the Royal Geographical Society, London, 6 June.

CLO (1994) *The Risonpalm Project Threatens the Ecosystem of the Niger Delta*, Civil Liberties Organisation, Lagos, Nigeria, September.

CoA (1992) 'Court of Auditors Annual Report for Financial Year 1991,' Official Journal of the European Communities, OJC C330, Vol. 35, Luxembourg, 15 December.

DANCED (1996) *Policy Guidelines for DANCED – Danish Cooperation for Environment and Development*, Ministry for Environment and Energy, Copenhagen.

DANIDA (1994) *Environmental Assessment for Sustainable Development 1994*, Ministry of Foreign Affairs – DANIDA, Copenhagen.

Dieterle, Gerhard (1996) 'Conservation of West Africa's Forests through Certification,' *The Courier*, No 157, May–June.

ECDPM (1996) *Beyond Lomé IV*, Policy Management Report No. 6, European Centre for Development Cooperation Management, Maastricht.

Ecooperation (1996) *Stichting Ecooperation*, Amsterdam.

EEB (1993) *Les ONG du Sud et la Coopération Européenné: Perception et Participation*, (Brussels: European Environmental Bureau), November.

Ewing, Winnifred (1996) 'The Results of UNCED – Development Aspects,' A3–1096, European Parliament, Brussels, March.

FOEE (1995) *Sustainable Europe* (Brussels: Friends of the Earth Europe).

Hayter, Teresa (1988) *Exploited Earth* (London: Earthscan).

ICPQL (1996) *Caring for the Future*, Independent Commission on Population and Quality of Life (Oxford: Oxford University Press).

IIED (1995) *Citizens' Action to Lighten Britain's Ecological Footprint* (London: International Institute for Environment and Development), February.

Jaycox, Edward (1993) 'Capacity Building: The Missing Link in African Development,' Speech to African-American Institute Conference, Virginia, May.

Karliner, Joshua (1994) 'The Environment Industry – Profiting from Pollution,' *The Ecologist*, Vol. 24, No 2, March/April, pp. 59–63.

Khosla, Ashok (1995) *Assistance for International Development Syndrome – Another Life-threatening Condition?* (New Delhi: Development Alternatives), October.

Milieudefensie (1992) *Sustainable Netherlands* (Amsterdam: Mileudefensie).

Myers, Norman (1995) *Environmental Exodus* (Washington, DC: Climate Institute), June.

NAR (1993) *The Environment – A Global Concern* (The Hague: National Advisory Council for Development Cooperation), June.

ODA (1991) *ATP Synthesis Evaluation Study* (London: Overseas Development Administration).

OECD (1994) *Export Promotion and Environmental Technologies* (Paris: Organisation for Economic Cooperation and Development Environment Directorate).

—— (1996a) *European Community*, Development Cooperation Review Series No. 12 (Paris: Organisation for Economic Cooperation and Development).

—— (1996b) *Development Cooperation 1995* (Paris: Organisation for Economic Cooperation and Development).

OTA (1993) *Development Assistance, Export Promotion and Environmental Technology* (Washington, DC: US Congress Office of Technology Assessment), August.

Putzel, James (1996) 'European Union Development Assistance: Transparency and Accountability,' Working Paper No. 96–01, Development Studies Institute, London School of Economics, London, May.

RF (1995) 'L'Action du Ministère de la Coopération dans le Secteur de L'environnement,' Ministère de la Coopération, République Française, Paris, April.

Rich, Bruce (1994) *Mortgaging the Earth* (London: Earthscan).

RIET (1996) *RIET in Focus*, Newsletter for the Regional Institute for Environmental Technology, Singapore, Vol. 2, No. 1, January.

Robins, Nick (1996a) 'Steering EU Development Cooperation Towards Sustainability: The Case of the Lomé Convention,' *European Environment*, Vol. 6, No. 1, pp. 1–5.

—— (1996b forthcoming) 'Greening European Union Foreign Policy', *European Environment*, Vol. 6, No. 6, pp. 173–182.

Sandbrook, Richard (1995) 'Steps to a Sustainable World – Progress Since the Earth Summit, International Institute for Environment and Development,' London, unpublished manuscript.

Santos, Ronet (1995) 'Transparency and Accountability of European Development Assistance,' Philippines Development Briefing, No. 9, Catholic Institute for International Relations, London, March.

SOLAGRAL (1996) 'Contradictions in European Policy Towards Developing Countries,' Solagral, Paris, January.

T&B Consult (1996) *Survey of Private Business and NGOs in Environmental Cooperation with Developing Countries*, DANCED, Ministry of Environment and Energy, Copenhagen.

UNDP (1993) *Human Development Report 1993*, United Nations Development Programme (Oxford: Oxford University Press).

WDM (1994) *Europe and the Third World* (London: World Development Movement).

Wenning, Marianne (1989) 'Background Papers for the Seminar to Promote a Coherent Ecological Approach in European Community–Third World Relations, Environment and Development Resource Centre, Brussels, April 1989.

Woodward, David and Jenny Pryke (1995) *The Common Agricultural Policy – Sustainable or Bankrupt?* (London: Catholic Institute for International Relations).

Worthington, Tony (1993) 'Our Millions Create Their Poverty,' *The Independent*, 7 June.

INDEX

acid rain 152
ACTIONAID 190, 191, 201
adaptation threshold 128
Agence Europe (AE) 49, 68, 148
Agrasot, P. 192
Agricultural Measures Compatible with
  the Requirements of the Protection
  of the Environment and the
  Maintenance of the Countryside,
  Council Regulation 2078/92 164
agriculture: domestic offsets 15;
  environmental degradation 163, 168;
  fertilizers 7; livestock effluents 163;
  pesticides 26; set-aside 164, 170;
  trade liberalisation 172–175; water
  pollution 165; *see also* Common
  Agricultural Policy
agro-chemical use 163
aid: economic globalisation 201;
  environmental 4, 21, 208; greening
  191–193; security 200; sustainable
  development 189–191
air conditioners 48
Allen, M. 85
ALTENER 14, 116, 118
American Forest and Paper Association
  151
Anderson, K. 120, 180
animal protection 147–148
Arden-Clarke, C. 7
Asia 126, 194–195
Aspinwall, M. 89, 91
Asteris, M. 89, 90, 91
Atochem 44, 47, 48, 50
Australia, prior informed consent
  procedure 76
Ausubel, J. 125

Baldock, D. 163, 164, 168, 169, 180
Bandarra, N. 180
Banki, W. 152
Barbera, A. 8
Bardot, Brigitte 147
Barnett, S. 116
Barrett, S. 8, 20
Bartik, T. 7
Basel Convention 75
BASF 61
Bayer 61
Beaufoy, G. 180
beef hormone ban 157
Belgium: carbon dioxide emissions 113;
  prior informed consent procedure 76;
  set-aside schemes 164
Benedick, Richard 39, 40, 42, 43, 44,
  45
Bergesen, H. 108, 109, 111, 122
Bergman, H. 24
Berlin Mandate 130
best available technology 6
Beuermann, C. 109
biotop disturbance 163
Birdlife International European
  Agriculture Task Force 180
BISD *see* GATT, Basic Instruments and
  Selected Documents
Bjerregaard, Ritt 49
Blazejcak, J. 7, 20
Blue Angel 150, 206
BNA International Trade Reporter 151,
  154
Bossuyt, J. 203
Box, Louk 200–201
Brazil: Carajas steel programme 192;
  ecolabelling 151–152